Forest Ecosystem in Modern World

Forest Ecosystem in Modern World

Pawan Pattanaik

RANDOM

RANDOM PUBLICATIONS

NEW DELHI (INDIA)

Forest Ecosystem in Modern World

ISBN 978-93-5111-829-9

Published in 2016 in India by

RANDOM PUBLICATIONS

4376-A/4B, Gali Murari Lal, Ansari Road
New Delhi-110 002
Phone : +9111-43580356, 011-23289044, 011-43142548
e-mail: sales@randompublications.com,
info@randompublications.com, randomexports@gmail.com

Reprinted 2026

Type Setting by : Friends Media, Delhi-110089

Preface

Forest ecosystems are areas of the landscape that are dominated by trees and consist of biologically integrated communities of plants, animals and microbes, together with the local soils (substrates) and atmospheres (climates) with which they interact. Forests are much more than the present population or community of trees. Forests that have been recently killed or altered by fire, insects, disease, wind or logging are still forests because of the biological and physical legacies from the previous forest - legacies of forest soil, organic matter, microbes, minor vegetation and animals. Under a regime of sustainable forest management, many or most of these legacies persist during the period between forest disturbance and the redevelopment of tree cover.

A forest ecosystem is one major ecologic unit that exists as "home" for a community of both native or introduced, classified organisms. The forest ecosystem is just one of a number of unique ecosystems including prairies, deserts, polar regions and great oceans, smaller lakes and rivers. A forest ecosystem typically is associated with land masses covered in trees and those trees are often classified by foresters into forest cover types.

A forest ecosystem community is directly related to species diversity. Generally, you can assume that the more complex the structure, the greater is its species diversity. You should remember that a forest community is much more than just the sum of its trees. A forest is a system that supports interacting units including trees, soil, insects, animals, and man.

The common idea for many people is that forests are just a collection of trees. However, they are much more than that. They are a complex, functional system of interacting and often interdependent biological, physical, and chemical components, the biological part of which has evolved to perpetuate itself. This complexity produces combinations of climate, soils, trees and plant species unique to each site, resulting in hundreds of different forest types around the world.

The problems of modern society are both complex and interdisciplinary.

– Author

Contents

1

Ecosystem Diversity in Forest

Analogous to species diversity, the number of ecosystems and pattern of their distribution can be used as a measure of ecosystem diversity. While we know of no examples where fishing activities eliminated an ecosystem, there are several examples where fishing activities have resulted in major reduction in the regional distribution of ecosystem types over large areas.

Mariculture—the farming of economically valuable sea life—is a fishing activity that has significantly altered coastal and estuarine habitats in many parts of the world. Along the coasts of Ecuador and Thailand, for example, fish farms have replaced mangrove habitats over fairly extensive areas.

In order to build aquaculture facilities to raise shrimp and fish, mangroves are dug out and replaced with ponds, eliminating essential nursery habitat for many fishes. While mangrove ecosystems have not disappeared on a global scale, on a regional level there have been significant reductions in the total area of mangroves; a form of ecosystem depletion.

There are other examples of ecosystem diversity being affected by fishing activities that destroy habitat upon which complex communities of marine organisms rely. Overfishing of herbivorous fishes on coral reef complexes in the Caribbean, as in the example above, has directly resulted in coral reef die-offs. The use of dynamite and cyanide in Southeast Asia to catch reef fish for local consumption and the aquarium trade has also killed off significant expanses of coral reefs. Trawling, where heavy nets are dragged along the ocean floor, can also damage seagrass or rocky habitats, physically dislocating or crushing fish and shellfish, undermining structural needs, and disrupting food availability for creatures such as shellfish and groundfish.

The dance of life operates in diverse, strange, and mysterious ways. Understanding the biological processes influencing the functional relationships of marine organisms, species, and whole communities has practical implications for understanding the consequence of human actions. As a species reliant on the biological productivity of oceans, in order to optimize the benefits humans can gain from this vessel of life, we need to know how to minimize the negative consequences of our actions for other living things, and for future generations.

Even with better understanding of the human factors influencing marine biological diversity, fisheries management is inevitably a matter of politics, not science. Management will not succeed in preserving future options unless it is both scientifically informed and ethically responsible.

MARINE BIODIVERSITY IMPORTANT

The seas provide a unique set of goods and services to society, including moderation of climate, processing of waste and toxicants, provision of vital food, medicines and employment for significant numbers of people. Our coasts provide space to live and directly and indirectly create wealth, including millions of jobs in industries such as fishing, aquaculture and tourism..

Looking at ecosystems in terms of the goods and services they provide allows us to realise their full value and our dependence on those systems in the broadest sense. Exploitation of the environment for one purpose can alter the environment's ability to provide other goods and services, so this knowledge is also a way of understanding what we stand to gain and lose by exploitation of certain aspects of the environment. The main goods and services provided by marine ecosystems are:

RESILIENCE AND RESISTANCE

Coastal and marine ecosystems are affected by environmental disturbance at a variety of spatio-temporal scales. The organisms inhabiting these systems are adapted to such disturbance, either by being tolerant of these conditions or by playing a role in one or more of the successional stages that follow during ecosystem recovery.

If all species in the system were tolerant to a particular perturbation, very little would change at the ecosystem level, and we could call the system resistant to this disturbance. However, often a disturbance, such as a temporary very low oxygen level, affects a substantial proportion of the organisms dramatically, either causing them to die, or forcing them to rapidly migrate to more favourable parts of the environment. Such a catastrophic disturbance could locally defaunate a certain volume in the pelagic or a certain area of hard or soft substrate.

Such destruction at a local scale does not mean the end of local functioning; Usually organisms are available at a larger spatial scale that can re-colonize the affected area, according to their particular tolerances and abilities to favourably affect their local environment. The term resilience has been used in different ways, first of all for the rate of recovery to the previous state and secondly for the system's ability to re-organize itself. Both resistance and resilience cause an ecosystem to remain relatively unchanged when confronted by a disturbance, but in the case of resistance no internal re-organization and successional change is needed. In contrast, resilience implies that the system is internally re-organizing, perhaps through a mozaic of patches that are at different stages of re-assembly. When considering the potential effect of a certain type of disturbance it is thus useful to ask two questions:

- Will the species of this system be able to tolerate it (implying resistance), and if not,
- Is recovery possible through a successional trajectory, back to the same, or at least a desirable, ecosystem state (implying resilience)?

It should be clear that the system will not be sufficiently resistant when (even gradual) uni-directional change acts faster than the organisms' ability to adapt their tolerances. If uni-directional gradual change is this fast, the system will not be sufficiently resilient either, as full recovery through succession will then not be possible. Recovery from sudden and local disturbance is usually possible through re-colonization, but the rate of recovery will depend tremendously on the spatial extent of disturbance. For example, recovery from anoxia could take 5 to 8 months at the scale of square meters, but could take 5 to 8 years at the scale of a whole bay.

Resilience has been defined in different ways: it can be a measure for the speed at which an ecosystem returns to its former state following a (minor) disturbance. The idea is that a system with a short return time is more resilient than one with a long return time. Such resilience measured as (1/the return time to a stable equilibrium) has also been called *engineering resilience*. It has however a long history of use among ecologists. It is also used in a way that more closely resembles the definition of resistance.

Biodiversity allows ecosystems to adapt to changing conditions. Humans, however, have acted to increase the rate of change and consequently, it will be a great challenge for the marine environment to adapt rapidly enough in the future. These changes have been induced through pollution, fishing, sediment deposition and alteration of the global climate. Without genetic diversity, natural selection cannot occur and natural selection is limited, then adaptation is impossible. It is evident that the preservation of biodiversity and, more specifically, genetic diversity is of paramount importance for successful adaptation to our rapidly changing environments.

Resilience through re-colonization

To understand resilience of ecosystems it is essential to understand what drives succession within these ecosystems. Succession determines how, and how fast, communities return back to their original state, or perhaps enter a new state. many aspects of succession can be understood in terms of trade-offs between the ability to be either a good early (re-)colonizer, or a good competitor (r-species versus a K-species). Succession involves a gradual replacement of species that differ in these traits and that differ in the degree they tolerate, facilitate or inhibit certain environmental conditions and other species.

We could thus also call a system resilient when it is organized in such a way that succession leads to a recovery of the original state.

Resilience and Nutrient Inputs

DeAngelis *et al.* 1989 showed in their analysis of a food chain model that resilience, measured as (1/return time), increased smoothly with nutrient enrichment. The idea is that the system can more quickly return to its original state when the nutrient input is high. However, this result depends on the exact shape of the functional response, the way predators consume their prey as a function of prey density. DeAngelis *et al.* had used a highly stabilizing Type III (a sigmoid-shaped) functional response. Such a functional response is realistic for learning predators and those that switch between different prey types. Vos *et al.* 2005 found that resilience (1/return time) would first increase, but then decrease again under nutrient enrichment, when a Type II functional response was used in the model. Many species of consumers throughout the animal kingdom consumer their resources with such a functional response, where intake rate first increases, and then gradually levels off with resource density. Food chains with such consumers would be much less resilient under a high pressure of nutrient enrichment.

RESISTANCE TO CHANGES IN ABIOTIC AND BIOTIC FACTORS

Community composition and ecosystem function may change very little under environmental change when the organisms can acclimate to such change or tolerate it for some time (when the change is only temporary). However, all organisms have bounds to what they can temporarily or permanently tolerate, and when change exceeds some of these limits, the community compostion and ecosystem functioning is likely to change. It is unlikely that communities can be resistant to continuous gradual change, such as global warming. Acclimation and phenotypic plasticity do not suffice to maintain the system as it is. Genetic adaptation could allow community members to track such abiotic environmental change, but it is more likely that the area where the community is functioning will be invaded by species that function well at higher temperatures. The original species will thus have to deal with new competitors and predators, in addition to the changed abiotic factor. To some extent the original community can track the preferred temperature range, by moving spatially to greater depths or to alternative gepgraphic areas. But these new areas are likely to differe in other ecological aspects such as water pressure, light climate and perhaps speeds of water flow etc.

Adaptation and the Consequences of Mortality

External disturbance interacts with internal mechanisms that shape community structure. To understand how an increased mortality of top-predators will affect the entire food chain, it is essential to understand how processes of mutual adaptation within food chains already give shape to existing patterns such as trophic structure (how biomass in ecosystems is partitioned between trophic levels (such as algae, herbivores, carnivores and top-predators)... Abundances at different trophic levels (such as algae, herbivores, carnivores and top-predators) and their responses to increased mortality (as under environmental change) depend critically on different mechanisms of adaptation within food chains and on the importance of density dependence at each of these trophic levels.

However, different types of adaptation to living in a food chain context (balancing the need to acquire resources with the need to avoid predation) can often have very similar consequences. For example, micro-evolution or behaviour, species replacement and induced defenses at a middle trophic level may all have similar effects on trophic level abundances in disturbed food chains.

Adaptation Assisted by Man

Protecting sources, not sinks when creating Marine Protected Areas. Protecting sources of populations at all stages of succession, to preserve 'ecological memory' to the fullest possible extent.

This includes protecting not only 'high quality' habitats that harbour healthy mature communities, but also 'low quality' and disturbed habitats that are required for those species that contribute to early recovery of perturbed areas.

Disturbance Prevention

Living marine flora and fauna can play a valuable role in the defence of coastal regions *i.e.* disturbance prevention. Marine ecosystems and the presence of organisms in the front line of sea defence can dampen and prevent the impact of tidal surges, storms and floods providing a 'buffering' effect that protects humans from the effects of these destructive perturbations. This disturbance alleviaion service is provided mainly by a diverse range of species which bind and stabilise sediments and create matural sea grass defences, for example salt marshes, mangrove forests and sea grass beds.

Sediments cover most of the seabed and hence most of the earth. Recycling of carbon and nutrients within this habitat (both subtidally and intertidally) is critical both at small and large scales. The availability of essential nutrients, such as nitrogen and phosphorus, and metals is essential for life. Processes that aid nutrient cycling are crucial to ecosystem functioning, as this increases the availability of nutrients and thus maintains productivity of the system. For example, in the marine benthic environment, bioturbation by marine worms, mainly through burrowing in the sediment, moves nutrients from deep sediment layers to the surface and vice versa. Nutrient cycling is also maintained through processes such as ingestion and excretion of materials by organisms *e.g.* fish mineralise nitrogen and phosphorus through excretion.

Gas and Climate Regulation

Marine ecosystems are an important regulator of global climate. Biogeochemical processes such as regulation of the CO_2/O_2 (carbon dioxide/ oxygen) balance, maintenance of the ozone (O_3) layer and sulphur oxides (SOx) are necessary to maintain a healthy planet and a healthy human population through the provision of breathable air. The biogeochemical cycling of gases is greatly controlled by the living biota existing on earth of which the marine realm is extremely important. For example, marine plants and animals aid in controlling carbon dioxide in the ocean, as phytoplankton remove it from the surface waters while releasing oxygen. When phytoplankton die, they sink and add to the supersaturation of carbon dioxide in the deep sea. This results in a vertical gradient of CO_2 in the ocean, which has been termed the 'Biological Pump.'

Greenhouse gas regulation is vital in regulating the climate of our planet. The seabed has a significant role in this process through its ability to sequester CO_2. Gases such as CO_2 in the atmosphere traps heat from the sun, heating the planet. This process occurs naturally and has kept the Earth's temperature about 60 degrees Fahrenheit warmer than it would otherwise be. Excessive amounts of gases such as atmospheric CO_2, however, can have a significant contribution to global warming and is thus a factor in regulating climate..

Bioremediation of Waste

A significant amount of human waste settles on the seafloor, through wash off from land and through transport from rivers and estuaries. Waste settling on the seabed is stored, assimilated, diluted and recycled through chemical re-composition, these bioremediation processes which de-toxify and purify waste and are of critical importance to the marine environment. Through either direct or indirect activity, marine living organisms store, bury and transform many waste materials through assimilation and chemical de and re-composition. For example, the bioturbation activity (reworking and mixing of sediments) of mega- and macrofaunal organisms within the seabed can bury, sequester and process waste material through assimilation and/or chemical alteration. Marine organisms such as the polychaete worm, Nereis succinea, takes heavy metals through feeding on sediments.

These detoxification and purification process are of critical importance to the health of the marine environment.

Biologically Mediated Habitat

Many organisms provide structured space or living habitat through their normal growth, for example, reef forming invertebrates, meadow forming sea grass beds and marine algae forests. These 'natural' marine habitats can provide an essential breeding and nursery space for plants and animals, which can be particularly important for the continued recruitment of commercial and/or subsistence species. Such a biologically mediated habitat can provide a refuge for plants and animals including surfaces for feeding and hiding places from predators. Living habitat plays a critical role in species interactions and regulation of population dynamics, and is a pre-requisite for the provision of many goods and services.

One of the most tangible services provided by the marine environment is the provision of Food for human consumption. Plants and animals derived directly from marine biodiversity provide a significant part of the human diet. Fisheries in particular, and the accompanying employment, provide a significant example of the importance of this function.

However, those species that are harvested commercially tend to be very heavily exploited, with stocks of many traditional favourites such as North Sea cod facing the threat of collapse. Seaweeds such as giant kelp, nori and agarweed are also very important commercially, historically being extremely important in East Asia. Conserving biodiversity of these groups will allow the use of under utilised resources in the future. A diverse ocean could potentially help to alleviate current and future commercial fishing pressures, provided these resources are managed responsibly.

Ornamental Resources

The oceans are a source of natural raw materials such as medicines, feed for livestock, polysaccharides and building materials. The potential for acquiring future raw materials from the marine environment is enormous.

Natural medicines continue to be discovered through the wealth of diversity on the planet. Although most medicines originate from sessile land plants, the ocean hosts many sessile animals that defend themselves through chemical means. The oceans host a high biochemical diversify resulting from the high phyletic diversity in marine ecosystems, suggesting there may be further potential to extract chemicals for pharmaceuticals from these organisms. An example includes the extract (arabinosides), collected from the sponge Tethya crypta and used in the treatment of herpes.

Red, brown and green algae provide a source of polysaccharides for a variety of human uses. Seaweeds are important in agriculture as feed for livestock and as compost for farmland. Chiton from shrimp and crab shells is used in agriculture as well as in human food supplements. An additional benefit of using these resources is that less of the animal is wasted from the fishery alone.

Aggregates extracted from the marine environment are used for building materials. Sand, gravel and crushed rock are used in the construction industry for housing and road building. Their removal has a physical impact on the seabed and habitat type through removal of material and resettlement of fine particles. Extraction must be carried out in a sustainable manner, as a healthy habitat is a prerequisite for the provision of goods and services from the marine environment. Marine biodiversity provides the basis for a wide range of leisure and recreational activities including: (sea) bird watching, rock pooling, beachcombing, sport fishing, recreational diving, and whale-watching. The provision of leisure and recreation results in significant employment opportunities.

Cultural Values

Cultural heritage and identity are intrinsically linked to the provision of food and employment is the support of cultural and spiritual traditions associated with fishing communities.

There is benefit associated with marine biodiversity for example for religion, folk lore, painting, cultural and spiritual traditions. Human communities living by and off the sea often attach special importance to marine ecosystems that have played a founding or significant role in the economic or cultural definition of the community. This identification may be associated with a strong economic interest in the extraction of the site but as economic significance decreases the community may attach increased symbolic values to the preservation of the site. For example a mussel bed may long have lost its economic significance while the symbolic importance may be high. This valuation should be distinguished from the economic importance of revitalised and commercialised cultural heritage which is included under the heading Leisure and recreation.

In Europe some of our most beautiful natural amenities are clean marine ecosystems, such as coasts, seas and estuaries, which enhance our landscapes and form an integral part of our cultural and natural heritage.

MANAGEMENT OF ECOSYSTEM

Practical ecosystem management usually involves maintaining a natural or artificial homeostasis in which the end product is of benefit to man, and in which violent or pathologic imbalances can be controlled. An example of a natural ecosystem which man might wish to maintain for esthetic or recreational value would be a wilderness area or National Park. Here the most important consideration is to insure that natural processes are permitted to continue. In a forest this would mean, for example, that fire, if already a natural part of the ecosystem, be permitted to occur at natural frequency and intensity.

Several recent studies have shown that certain forests depend upon intermittent burning to recycle nutrients and minerals. In the lowland conifer forests of Alaska, if fire is excluded a thick carpet of moss may accumulate and raise the permafrost level. This encourages the growth of black spruce, a species with little food or timber value. Browsing animals such as deer, moose and bear cannot survive on black spruce. They depend on the Tarai fires to maintain a suitable cycle of vegetation. Many foresters now feel that the underbrush and leaf litter in the Ponderosa pine forests of western United States should be removed by light ground fires every few years to maintain soil fertility, to prevent excessive accumulation of organic matter in the form of cellulose and also to prevent more devastating fires that destroy the entire forest.

In the pine forests of the South, very efficient fire suppression techniques allowed dangerous accumulation of combustible materials, and a pine fungus disease was favoured in the accumulated litter. In all of these forests, proper management should not exclude all fires, but should prevent totally devastating fires. These examples illustrate the need for understanding natural processes in ecosystems in order to maintain them properly.

The Everglades of Florida provide another example of the importance of understanding natural ecosystem processes before the system can be properly managed or maintained. The unique flora and fauna of the Everglades depends upon the proper balance of fresh and saline water flowing through its channels. If a significant decline in fresh water input occurs, as has been taking place due to the diversion of fresh water to the growing megalopolis of Miami, numerous important species of plants and animals may die and the characteristic features of the system may be destroyed. In times of drought, many animals depend upon alligator wallows as a source of water. These are pools hollowed out by alligators, and they tend to retain water longer than other parts of the swamp. If alligators decline through shooting or any other.

It is increasingly important to understand the dynamics of ecosystem, their patterns of succession and evolution, and their patterns of succession and evolution, and their patterns of diversity and stability. E.P. Odum has written a major review of these topics and has shown that ecosystem may be characterized as pioneer, developmental and mature. Various stages within this continuum possess certain properties of production, respiration, community structure, nutrient cycling, homeostasis, etc. For example, young developmental stages of an ecosystem tend to have greater gross production in relation to respiration, but lower species diversity and less stability than mature stages.

Thus, developmental stage to be favoured in agricultural practice, but they are also more vulnerable to ecologic damage or catastrophe. We will explore these relationships in more detail throughout subsequent chapters after various components of the ecosystem are considered seperately, diversity and stability. E.P. Odum has written a major review of these topics and has shown that ecosystem may be characterized as pioneer, developmental and mature.

Various stages within this continuum possess certain properties of production, respiration, community structure, nutrient cycling, homeostasis, etc. For example, young developmental stages of an ecosystem tend to have greater gross production in relation to respiration, but lower species diversity and less stability than mature stages. Thus, developmental stage to be favoured in agricultural practice, but they are also more vulnerable to ecologic damage or catastrophe. We will explore these relationships in more detail throughout subsequent chapters after various components of the ecosystem are considered seperately.

THE PRINCIPLES OF ECOSYSTEM MANAGEMENT

Even if some synergies between the different international arrangements and processes relating to the biodiversity of freshwater resources still deserve more attention and none of the international treaties and institutions already deals fully with freshwater ecosystems, there is an increasing consensus on the highest political levels to manage water, water-related processes, and ecosystems in a sustainable manner.

However, at the level of implementation the value of ecosystems is rarely taken fully into account. In order to close this implementation gap the contracting parties of the CBD and other advocates have proposed a so-called "ecosystem-based approach" as an appropriate way of managing natural resources.

THE ECOSYSTEM APPROACH

The ecosystem approach is a strategy for the integrated management of land, water and living resources that promotes conservation and sustainable use in an equitable way. The term ecosystem is used more as a mental construct suggesting complexity and systems interaction rather than a geographic entity. An emphasis on the complexity of system-wide interactions highlights scientific uncertainty and results in the need to deal with this uncertainty explicitly by acting conservatively and managing adaptively: setting a course of action based on a set of hypotheses, monitoring what happens, and re-evaluating the direction on what one learns.

At its most fundamental level, an ecosystem approach maintains diversity as a means of building resilience against catastrophic events in biological, economic, organisational, and political systems. At its second meeting in 1995, the Conference of the Parties of the CBD adopted the ecosystem approach as the primary framework for action, and subsequently has referred to the ecosystem approach in the elaboration and implementation of the various thematic and cross-cutting issues work programmes under the Convention. The issues concerned include:

- Biological diversity of inland water ecosystems.
- Marine and coastal biological diversity.
- Agricultural biological diversity.
- Forest biological diversity.
- Indicators of biological diversity.
- Incentive measures and environmental impact assessment.

By recommending the ecosystem approach as the guiding concept, in May 2000 the fifth COP took a next step towards its practical verification in the frame of the CBD. Concerning inland water ecosystems a work programme was adopted by the COP which highlights the importance of designing integrated watershed, catchment and river basin management strategies. Particularly the interrelation between water management and agriculture was stressed, as well as the need to use the ecosystem approach in all sectors which have an impact on inland water biodiversity.

The ecosystem approach was initially developed as a research paradigm and not intended to serve as the basis for resource management. With the political application of the notion "ecosystem-based approach" advocates intended to find an adequate language in response to the shortcomings of classical nature conservation approaches.

Although classical nature conservation (*e.g.* designation of protected areas) can be considered successful in many regions and also is an important module in ecosystem management, its recognition of human needs and other legitimate sectoral interests was rather insufficient. It did not put enough emphasis on the biological diversity outside protected areas. Furthermore, the primary method of protecting freshwater biodiversity has been to designate particular species as threatened or endangered making them subject to national recovery programmes or international protection.

Unfortunately, this approach is failing and it demonstrates the need for a more holistic approach in order to overcome the shortcomings of traditional efforts in nature protection policy. In the United States, for example, no aquatic species has ever graduated from the government's endangered species list, but 10 species of fish have been removed due to extinction. Equally, ecosystem management responds to the obvious shortcomings of classical single-resource approaches focused on exploitation.

THE 12 MALAWI PRINCIPLES

- Management objectives are a matter of societal choice.
- Management should be decentralised to the lowest appropriate level.
- Ecosystem managers should consider the effects (actual or potential) of their activities on adjacent and other ecosystems.
- Recognising potential gains from management there is a need to understand the ecosystem in an economic context. Any ecosystem management programme should:
  - Reduce those market distortions that adversely affect biological diversity.
  - Align incentives to promote sustainable use.
  - Internalise costs and benefits in the given ecosystem to the extent feasible.

- A key feature of the ecosystem approach includes conservation of ecosystem structure and functioning.
- Ecosystems must be managed within the limits to their functioning.
- The ecosystem approach should be undertaken at the appropriate scale.
- Recognising the varying temporal scales and lag effects which characterise ecosystem processes, objectives for ecosystem management should be set for the long term.
- Management must recognise that change is inevitable.
- The ecosystem approach should seek the appropriate balance between conservation and use of biological diversity.
- The ecosystem approach should consider all forms of relevant information, including scientific and indigenous and local knowledge, innovations and practices.
- The ecosystem approach should involve all relevant sectors of society and scientific disciplines. These principles were agreed at CBD's COP-5 under decision V/6 with minor modifications.

Therefore, ecosystem management is proposed as a modern way of managing natural systems and generally intends to use natural resources within their natural limits. However, although there are still some controversies on how to translate ecosystem management into concrete management terms, parties of the CBD have already agreed on basic principles to guide the *modus operandi*. A key point is that the ecosystem approach is a holistic process for integrating and delivering in a balanced way the three objectives of the CBD: conservation and sustainable use of biological diversity combined with equitable sharing of biodiversity's benefits. In particular, the agreed principles of the ecosystem approach stress the importance of:

- Societal choice concerning management objectives.
- The "economics" of biodiversity (internalisation of costs and benefits, balance between conservation and use).
- Adequate institutional structures (decentralisation, management at the appropriate scale).
- Adaptive management which enables policy makers to anticipate and cater for the natural dynamics of ecosystems.
- Consideration not only of scientific knowledge but also of indigenous and local knowledge.

In order to translate this rather abstract principles into guidelines for water management it is first necessary to define the appropriate management level to meet the general objectives. While the cited principles stress the importance of decentralised approaches in general terms, the idea of river basin management is considered an adequate starting point for the implementation of the ecosystem approach in water management.

The reasoning is that river basins offer many advantages for strategic planning and that they represent the most appropriate geomorphologic units on which to base ecosystem management practices.

However, concerning ecosystem protection there are some limits: groundwater catchments only rarely match with surface water catchments; water quality and quantity inside the river basin interacts with other influences outside the catchments boundaries; estuarine ecosystems, for example, can only be sufficiently protected by taking into account the interaction of freshwater and tidal currents.

Therefore, river basin management should not be perceived as a panacea but it does provide a sound geographical basis for implementation. Briefly, ecosystem management should be an integral component of comprehensive river basin management strategies. Hence, the explicit switch to ecosystem management leads to an accentuation of the main focus in water management because ecosystem management demands:

- A broader perspective: whole ecosystems and the full range of services, no sectoral view.
- Efforts to balance natural ecosystem robustness and human-induced alterations.
- Collation of information on the status of, and threats to, the biodiversity of the freshwater ecosystem concerned (including impacts of the introduction of alien species).
- Explicit consideration of land-use, in-stream flow and water temperature for maintaining ecological functions.
- A long-term perspective.
- Consideration of the full economic value of freshwater ecosystems (include its biodiversity).
- Recognition of ecosystem's change as a matter of fact (*i.e.* restoration of damaged or destroyed ecosystems will never lead to the "pristine" state; adaptive management to deal with uncertainties).
- Integration of scientific knowledge and modern management tools (Decision Support Systems, Geographical Information Systems etc.) into decision-making.
- Identification of win-win options between protection and use of freshwater ecosystems: "use them, but don't lose them".
- Consideration of the full range of policy instruments applicable in water management (economic and legal instrument, co-operation etc.)

As most of these elements are widely agreed upon the rhetoric of international water policy, the next steps must include attempts to tackle the implications of the ecosystem approach in practice. For example, only little is known about the value of ecosystem management in political terms for at least two reasons:

- The concept appears highly demanding because much information – sometimes on "invisible" relationships between human-induced pressures and natural reactions – is needed, and effective vertical and horizontal co-ordination of sector policies and planning instruments has to take place. Today, even in wealthy nations a lack of integration across sectors in managing water resources can be identified.
- It is necessary to decide which services of water receive priority, and how the costs and benefits are to be distributed. The developed world, for example, can "afford" to stop or to remove development projects in favour of ecosystem integrity (*e.g.* Great Whale Hydropower Project in Canada, Edwards Dam in the US) and to use sometimes generous subsidies to compensate land owners for restricting their use of the lands. The restoration of river ecosystems in the developed world also is an example for costs emerging in the short terms. Although long-term economic benefits are regularly provided, political decision are often dominated by short-term considerations.

CONSISTING OF ECOSYSTEMS

It is helpful to visualize ecosystems as consisting of four basic components: abiotic substances, producer organisms, consumer organisms and decomposer organisms.

Abiotic Components

Abiotic substances are the inorganic and organic substances not momentarily present in living organisms. These include water, carbon dioxide, oxygen, nitrogen, minerals, salts, acids, bases and the entire range of elements and compounds outside living organisms at any given point in time. Many elements may be tightly bound in inorganic compounds, such as silicon in sandstone or aluminum in feldspar, and are unavailable to living organisms. Elements which are readily available to living organisms such as free O_2 or CO_2, or they may be in an inaccessible form such as silicon dioxide (SiO_2) in quartz, a major component of granite. Similarly, potassium may be readily available to plants in the form of KCL in soil, but relatively unavailable in the form of $KAlSi_3O_8$ in orthoclase or monoclinic feldspar, one of the commonest of all minerals.

An important property of an ecosystem which determines its productivity is the form and composition in which bioactive elements and compounds occur. For example, an ecosystem may have a substantial abundance of vital nutrients, such as nitrates and phosphates, but if they are present in relatively insoluble particulate form as they would be if linked to ferric ions, they would not be so readily available to plants as if they were in the soluble form of potassium or calcium nitrate and phosphate. One of the most important qualities of an ecosystem is the rate of release of nutrients from solids, for this regulates the rate of function of the entire system.

Producers

Producer organisms are bacteria and plants which synthesize organic compounds. They are said to be autotrophic or self-productive, in that they take inorganic compounds and manufacture organic materials and living protoplasm from them. All green plants, including microscopic algae, are producer organisms since they exhibit photosynthesis, and some bacteria are producers since they may exhibit chemosynthesis or photosynthesis. Obviously, all life depends upon the basic productive capacity of green plants and bacteria.

Consumers

Consumer organisms are animals which utilize the organic materials directly or indirectly manufactured by plants. Consumers are unable to produce their own organic compounds for basic nutritive purposes. They are said to be heterotrophic, which means different or varied in nutritional source. Primary consumers or herbivores directly consume the organic compounds of plants. Secondary consumers may be omnivores or carnivores which depend partially or entirely on other animals for food. Tertiary and quartenary consumers may be the second or third-stage predator, for example, a hawk feeding on a weasel which in turn consumed a mouse.

Decomposers

Decomposer organisms are bacteria and fungi which degrade organic compounds. Their nutrition is said to be saprophytic, that is, associated with rotten an ecosystem—they reduce the complex organic molecules of dead plants and animals to simpler organic compounds which can be absorbed by green plants as vital nutrients. They provide the final essential link in the cycle of life. They are necessary for the renewal of life, for if decomposers were not active, organic compounds would become locked into complex insoluble molecules which could not be utilized as nutrients by plants.

Ecosystems involve, of course, a wide variety of life forms not specifically mentioned in the preceding paragraphs, but virtually all components of an ecosystem can be classified into producers, consumers, or decomposers. For example, parasites are merely specialized consumers. Plant parasites feed directly on plants and are thus herbivores; animal parasites derive their nutrition from other animals, and are thus carnivores differing from predators only in the fact that they normally do not kill the host. Scavengers such as vultures are also carnivores, differing from predators by the fact that they feed on an animal after it has died from some other cause. Figures 1-2 and 1-3 illustrate the essential components of ecosystems and their relationships.

NATIONAL INSTITUTIONS AND ECOSYSTEM MANAGEMENT

A key prerequisite for successful implementation of an ecosystem-based approach in river basin management are adequate institutions that reach the full range of uses and users that are connected in freshwater ecosystems. While the need for integrated and cross-disciplinary approaches to land and water management are sufficiently described in theory, too little is known how the demanding concept could succeed in practice.

Local/regional Government Boundaries

Freshwater systems extend over a wide range of administrative and bureaucratic boundaries. It is rare that a single management unit has the full responsibility for the management of a freshwater ecosystem. Therefore, the failure to achieve integrated and environmental-orientated management has been attributed to the strength of local governments and sectoral orientated administrations. This is obviously confirmed by many cases of sector-by-sector fragmented approaches which remain the norm.

Today, there are at least some encouraging examples that prove that ecosystem management of water resources can be fostered by political measures. The new approach of the European Union (Water Framework Directive) is a frequently cited example for an institutional innovation that will help to strengthen the integration of ecological objectives in water management, basin by basin. Obligations for the member states of the European Union not only encompass the establishment of integrated river basin planning as a key instrument but also the consideration of the effects of water use on aquatic and related terrestrial ecosystems.

Although there is still a main focus on water quality, habitats and protection areas, for example, are an important part of the management plans to develop. Equally, the Tennessee River Basin, the new Brazilian water law and the new South African water law show promise for implementing an ecosystem-orientated management.

THE GREAT LAKES EXPERIENCE IN ECOSYSTEM MANAGEMENT

The Great lakes basin is a significant international ecosystem. Shared between Canada and the United States, it contains approximately 20 per cent of the Earth's fresh surface water and is home to some 35 million people. Environmental pressure is also a characteristic of the Great Lakes, *i.e.* eutrophication, contamination from toxic substances, fluctuating water levels, introduction of alien species, loss of native fish species and wetlands.

Despite still existing pollution and deficits in implementation in the two countries (different paces, partly non-compliance), Canada and the US have created an effective institutional arrangement for implementation of an ecosystem approach: The International Joint Commission - a longstanding, binational, generalist organisation assisted by several newer binational bodies with different tasks.

Substantial progress was made towards ecological goals. Further, co-operation of Governments has changed behaviour towards the entire ecosystem. Significant resources were devoted to studying the ecosystem and to carrying out clean-up programmes and restoration measures. Collaboration with industries was effective and broad participation (states and provinces, environmental and citizen organisations) was ensured. Despite serious disagreements at times, the two countries as well as the sub-national political entities have continued to participate in the Joint Commission.

However, it is worth mentioning that international co-operation in comprehensive river basin management usually takes a very long time to develop and often starts with less demanding arrangements than a more holistic ecosystem approach. The establishment of the International Commission for the Protection of the Rhine (ICPR) took place in 1950, but common measures to reduce the increasing pollution of the river were not agreed upon until the 1970s.

Further, much progress was achieved in the aftermath of the Sandoz disaster in 1986. In reaction to the disaster riparian states agreed on the Rhine Action Plan, a comprehensive plan to tackle environmental problems in a holistic manner including water quality aspects and the protection and restoration of habitats and floodplains. Further, in 1998 riparian states adopted an Action Plan for Flood Defence which includes, apart from water management, measures in the fields of nature protection policy, agriculture and forestry and spatial planning.

Today, the ICPR serves as an example for good practices in ecosystem management of international river basin because it explicitly considers biodiversity aspects and the interaction of land-use development and water management. However, one key lesson from the ICPR is that non-legally binding action plans are worth considering in international co-operation because they can be prepared more quickly and can be more ambitious as legally binding treaties.

The co-operation on the River Rhine has also shown that the identification of a flagship species can serve as a symbol for ecosystem management and might help to mobilise public support, *i.e.* efforts to develop a comprehensive ecological management approach were heavily stimulated by the programme *Salmon 2000* started in 1991. However, it was the combination of the Action Plans, the emergence of mutual trust, owing to the co-operative atmosphere in formal organisational structures, and the provision of considerable financial resources which made the ICPR a success story.

There are also some encouraging examples for improved consideration of environmental and biodiversity aspects in the developing world, even if the label "ecosystem approach" is rarely applied. The Mahakali Treaty, for example, explicitly addresses issues of river ecosystem maintenance in Uttar Pradesh. Even if economic aspects of water use and sharing of benefits are dominant, co-operation between India and Nepal will also lead to an amelioration of the river basin in environmental terms.

Further, the conflict between India and Bangladesh exemplifies the possibility for common action to tackle the problem of over-extraction leading to reduced availability of water resource in the downstream riparian state. In that case, a barrage was built 18 kilometres from the Bangladesh border on the Ganges River to divert water through a feeder canal to supplement the dry season flow of the Bhagirathi-Hooghly, the river that serves Calcutta and irrigation needs in West Bengal. On occasion this had reduced the flow of the Ganges substantially, causing damage to the riparian environment, and in particular affecting the world's largest mangrove forest, the Sunderabans.

A number of attempts had been made to resolve this dispute, culminating in the signing of the treaty between the two countries in 1997 that defined the mechanisms for sharing the Ganges waters. Other examples demonstrate that the development of a co-ordinated water resource development and management scheme is crucial. This is often the major aim of harnessing riparian states, as borne out by many cases in the past. A realistic basin-wide management scheme may be developed only through the collaboration of riparian states.

Having such a scheme serves as proof for other countries, in particular donor countries and aid agencies, that riparian countries are on good terms, to the extent that they have developed the scheme through collaboration and are ready to implement it in a co-ordinated manner. An example is the development scheme of the Mekong River Basin in South- East Asia, although many problems could not be solved. The management of the Okavango Delta and its biodiversity rich wetlands also has improved albeit increased water withdrawals for urban use is still a politically controversial issue and has threatened significantly biodiversity in the past.

Further, the Lake Chad Basin countries are actually undertaking a project for the sustainable management of groundwater, wetland and land resources in the basin. In addition, transboundary conservation areas (TBCAs) provide a mechanism to promote sound ecological management of transborder ecosystems, and, at the same, promote opportunities for regional political, economic, and cultural co-operation. Ecological benefits of TBCAs are mainly associated with protecting larger areas. Specifically, TBCAs may improve the protection of internationally shared resources, such as watersheds, that provide important ecological and economic services.

TBCAs may encompass critical portions of watersheds for multiple purposes, such as pollution control, clean water supply, wetland management, and carbon sinks. Hence, TBCAs might serve as a core zone for integrated management in entire river basin. Recently, such transfrontier conservation projects are taking place on East African Lakes (*e.g.* Lake Victoria, Lake Malawi), for example, which are globally important for biodiversity as centres of evolution for African freshwater fish.

Similar actions are occurring between South Africa and its neighbouring countries. In all these cases, joint international recognition of transboundary conservation projects, such as World Heritage Sites and Peace Parks, might strengthen co-operation. Because many of the approaches and projects mentioned are still in their early stages, it is impossible to judge their success in terms of ecosystem management. But even if progress can be expected, rapid changes will rarely occur because of the lack of adequate institutional capacities in many countries concerned.

This means that international co-operation might favour more comprehensive and ecosystem-based approaches in water management, but as vulnerability and degradation of freshwater ecosystems are mostly rooted in incomplete or inappropriate legal and policy frameworks at the national level the pace in implementing ecosystem-based strategies heavily depends on domestic policy reforms. Different stages of economic development and diverging "philosophies" in water management (*e.g.* water rights, pricing) make ecosystem management a long-term and complicated process.

As success stories have proven effective ecosystem management of transboundary freshwater ecosystems is usually part of a broader evolutionary process of building capacity towards co-operative water resources frameworks.

BIOLOGICAL DIVERSITY CONSERVATION FUNCTIONS

There is no strict divide between "wild" and "domesticated" species important for food and livelihoods. Many wild plant species and populations that have been considered to be wild are in fact carefully nurtured by people. A similar continuum exists for animal species that use agroecosystems as habitat, nesting grounds and food. Whilst not necessarily the subject of conscious management by herders or farmers, many wild species thrive in, or are dependent on, agroecosystems. In general the more structurally and biologically complex the agroecosystems, the more diverse the forms of wildlife. For example:

- The systems of shifting cultivation common in Asia have long been considered to be damaging to biological diversity in particular. However, given the opportunity, shifting cultivators preserve wild resources. Studies in Asia have concluded that most of the mature forests in this region are not virgin forests, but merely old forests that have reached a relatively stable dynamic state after some earlier clearing by human or natural means. Under traditional systems of shifting cultivation which speed up natural succession, wildlife flourishes, with elephants, wild cattle, deer, and wild pigs all feeding in the abandoned fields. Tigers, leopards and other predators are in turn attracted by the herbivores. Older fields contain a high proportion of fruit trees that attract primates, hornbills, squirrels and a variety of other animals.

- Long term research in the Lowland of west Poland shows that a mosaic of landscape structure of small cultivated fields, shelterbelts, meadows and small ponds helps maintain biological diversity (and many other functions such as enhanced water storage). Mammal, amphibian, reptile, insect diversity are enhanced and the number of bird species is positively correlated with the diversity index of the landscape.

Climate Functions

Although agricultural and the atmosphere interact in many ways, the links between climate and weather, and agricultural biodiversity can be rather complex.

At the global scale, the distribution of individual plants and animals, vegetation and crops is conditioned by available climatic resources such as solar radiation, which controls the production potential, and by rainfall, which determines to what extent the radiant energy can actually be used by plants for their growth. Indeed, it is for these reasons that climatic classifications largely coincide with vegetation maps.

At the local scale, types of landscapes and vegetation, including crop and crop-vegetation mixes contribute towards modifying the local climates by directly affecting wind patterns, rates of evaporation, rainfall interception (effective rainfall), etc. This in turn conditions the development of vegetation and crop canopies which can be said to create their own microclimate, resulting from the interaction of plants with the general climate. Shelter belts, or the use or large tree belts to protect tropical plantation crops from cyclones, provide clear examples in which people and communities derive direct advantage from these landscape features.

Many examples demonstrate the benefits of maintaining minimum agrobiodiversity in the face of climate variability. Recent droughts in some southern African countries for example, have shown that mixes of local varieties planted over several weeks have the potential to better resist unusual patterns of rainfall variability (*e.g.* early-season drought) than some modern varieties planted on the "optimum dates". In fact, there appears to be a link between crop biodiversity and relatively low but regular production, one of the keys to food security. Similarly, atmospheric pollution has been shown to interact with agro-ecosystem functions, both positively (nutrients, such as nitrates and sulphur) and negatively (toxic compound like tropospheric ozone and heavy metals). Given that the response to these substances varies enormously across the spectrum of species, they have the potential to modify patterns of biological diversity.

Agriculture also contributes towards the emission of some of the most significant "greenhouse" gases (in terms of global warming potential, GWP). Some studies have suggested for example, that increases in the area of permanently or quasi-permanently flooded rice cultivation account for a significant proportion of the increase in net methane emissions. The increased specialisation of ruminant production based on high yielding breeds has also significantly contributed to global methane emissions (methane is produced by anaerobic digestion in animals).

An extreme case would be the enhanced biogenic emissions of carbonylsulphide associated with increasing cultivation of high sulphur crops such as rape. Oxidation of carbonylsulphide in the stratosphere leads to the production of sulphate aerosols that influence the intensity of ultra violet radiation reaching the earth's surface, potentially affecting the dynamic functions of agricultural biodiversity. Volatile organic compounds such as terpenes and isoprenes are produced and released into the atmosphere by many plant species, especially in agroecosystems dominated by Mediterranean shrubs, eucalyptus and conifers.

By influencing the oxidation capacity of the troposphere, these volatile compounds influence the abundance and distribution of trace gases such as ozone. In almost all cases however, many of the these effects are marginal when compared with the climatic effects of land use change, including deforestation, which have until recently constituted one of the main non-industrial sources of carbon dioxide

Functions in the Water Cycle

There are few experimental studies exploring the links between agricultural biodiversity and water. However, available evidence shows that agricultural biodiversity plays a crucial role in cycling water from the soil to the atmosphere and back.

It also has measurable impacts on water quality. Agroecosystems with different species assemblages and plant architectures result in differences in the amount of precipitation intercepted, the proportion of precipitation converted to stem flow, and the proportion of precipitation that infiltrates the soil rather than running off. At the landscape level, the types, relative abundances and relative spatial locations of agro-ecosystem types affect the amount of water moving from one point to another. For example, conversion of vegetation within a watershed from forest or shrub land to less structurally and biologically complex grassland is known to influence stream flow out of the watershed, in both temperate and tropical systems.

At the functional group level, the root structure, phenology and physiology of different plant species important for food and agriculture have direct implications for the quantity and timing of water transfer to the atmosphere *via* evapotranspiration. Individual plant species differ in their resistance to water stress, their efficiency of water uptake from the soil, and so on. Genetic variations among crop varieties (differences in water use efficiency, stomatal resistance...) also influence these processes. At the landscape level, evapotranspiration from agroecosystems can have an effect on relative humidity and microclimate downwind.

In both terrestrial and aquatic environments, the plant and animal components of agricultural biodiversity also function to alter water quality by performing various filtration, uptake and excretory processes. These functions all affect the composition and concentration of dissolved gases, solutes and particulates. Species level differences in physiology can positively or negatively affect water quality. However, in most cases a greater diversity of biological organisms (from microbes to fish and macrophytes) leads to a higher quality of water for human consumption and use.

Land Use: Environment and the Sustainability

Land-use changes can affect the environment and the sustainability of production. Because impacts on the environment—including erosion, water quality, and wildlife habitat—are typically not reflected in private profit calculations, land-use choices that are optimal for an individual may not be optimal for society.

This difference suggests the possibility of public policies that more closely align land-use decisions with social objectives. The allocation of a fixed land base among competing uses is determined by the relative returns to the different uses, which vary according to land quality and location. A landowner seeking to maximize profits will allocate a land parcel to the use that yields the highest expected economic return, after the costs of conversion. As relative returns change along with market conditions, technological advancements, or government policies, land-use patterns tend to adjust accordingly. Land-use change is dynamic.

With the exception of urban land, changes occur to and from major land uses. For example, 44 million acres left the cropland and pasture category from 1992 to 1997 while 21 million acres shifted into the category, resulting in a net loss of 23 million acres.

Forest-use Land

Table. Major uses of land, United States, 2002[1]

Land use	48 States Million acres	US	48 States % of total	US
Cropland[2]	441	442	23.3	19.5
Grassland pasture and range	584	587	30.8	25.9
Forest-use land[3]	559	651	29.5	28.8
Special uses[4]	153	297	8.1	13.1
Urban	59	60	3.1	2.6
Miscellaneous other land	97	228	5.1	10.1
Total land area[5]	1,894	2,264	100.0	100.0

Note:

[1]See Major Land Uses for estimates of major uses by region and State, coinciding with each census of agriculture, from 1945 through 2002.

[2]All land in the crop rotation (used for crops, used for pasture, idle cropland). Includes about 34 million acres idled under the Conservation Reserve Programme.

[3]Total forest land as classified by the U.S. Forest Service minus an estimated 98 million acres of forested land used for parks, wildlife areas, and other special uses.

[4]Rural transportation areas, land used primarily for recreation and wildlife purposes, various public installations and facilities, farmsteads, and farm roads/lanes. Excludes urban land in contrast to Major Land Uses, Aggregate Data.

[5]Distributions by major use may not add to totals due to rounding.

Forest-use land, the second largest major use, declined from about 32 percent of total land in 1945 to about 30 percent in 2002. A broader category, all land with forest cover, comprised 33 percent of the land base in 2002. While forest-use land increased 1 percent between 1997 and 2002, it declined from 612 million acres in 1964 to 559 million acres in 2002. Much forest-covered land is in "special uses" that prohibit forestry uses such as timber production. Forested land in these special uses increased from 23 million acres in 1945 to about 98 million acres in 2002.

Cropland comprises the third largest use of land, covering 23 percent of the contiguous States in 2002. Since 1945, cropland ranged from a high of 478 million acres in 1949 to a low of 441 million acres in 2002. Total cropland has trended downward since the late 1960s, and decreased by 13 million acres (3 percent) from 1997 to 2002. The total cropland base includes cropland used for crops, cropland used for pasture, and cropland idled. These components vary more than total cropland. Since 1945, the amount of cropland used for crops has ranged fromas much as 383 million acres in 1949 and 1982 to a minimum of 331 million acres in 1987. Total acreage used for crops exhibited two major cycles between 1945 and 1987, with cropland moving from idle to crop use and back again. Cropland used for crops increased from 331 to 349 million acres over 1987-97, and then declined to 340 million acres in 2002, about 5 percent below the average acreage for 1910-97. Since 1945, cropland used for pasture varied from 47 million acres in 1945 to 88 million acres in 1969.

Major Land Uses

This series contains acreage estimates of major uses by region and State, coinciding with each census of agriculture from 1945 through 2002. Because Alaska and Hawaii have very little crop area, we focus on the contiguous 48 States. The total land area of the 48 contiguous States is approximately 1.9 billion acres, with an additional 365 million acres in Alaska and a little over 4 million acres in Hawai.

Grassland pasture and range, the largest use of land, accounted for 584 million acres (31 percent) of the 48 States in 2002. This compares with 636 million acres in the mid-1960s. Due to improvements in the forage quality and productivity of grazing lands, less pasture and range is needed to sustain grazing herds. The inventory of domestic animals, particularly sheep, has also been declining in recent years, further reducing pasture/range demand.

Land Alienation and Pauperization

The colonial state imposed the land laws based on this worldview on the traditional communities. This imposition affected the dominant castes as well as the tribal communities. But many of the former had access to education and other modern inputs. So they had some preparation to deal with the changes. Most traditional tribes, on the contrary, lived on mineral and forest rich land that the colonialist required as raw materials. That turned the imposition of the formal system on the informal societies into an unequal encounter. Land alienation from the traditional to "modern" communities was a consequences since the latter were unable to deal with the changes imposed on them.

This unequal encounter continues to be the basis of a disjunction and of conflicts between the two systems because the colonial laws continue to be in force in the country. One of its consequences is environmental degradation. The legal system that recognises only individual ownership is a major cause of land loss and environmental degradation. Since the CPRS are not recognised as their sustenance, the communities depending on them cannot prevent outsiders from encroaching on that land. For example, in Tripura in North Eastern India, the tribal proportion has declined from 58 percent in 1951 to 31 percent in 2001 because immigrants have encroached on 60 percent of their community owned land with the help of individual-based laws. Equally important is loss of forests which catered to many needs of the tribal and other rural poor communities. The state handed many of them over to industry as raw material. They were treated as sources of profit and destroyed with no concern for their dependants or for conservation. That impoverished people (Gadgil 1989.

The third source of land loss is acquisition for development projects. The law that empowers the state to acquire land recognises only individual ownership. More than 25 million hectares have been used in India for such projects 1947-2000, around 14 million of them forest and other CPRs. Their inhabitants, most of them tribal and other rural poor like fish and quarry workers are considered encroachers and are not compensated and often not even counted among the displaced (Fernandes 2008: 92).

Often records of the CPRs are not kept since they are considered state property and their inhabitants are encroachers. For example, according to official accounts, in Assam the state used 159,017.37 hectares of land for development projects and displaced 343,262 persons from them 1947–2000. The reality is 1.9 million persons displaced from 567,281.29 hectares (Fernandes and Bharali 2006: 107). More than 1.5 millions displaced persons and 410,261 hectares were not counted because according to the law these CPRs are state property and their inhabitants are encroachers with no right to live there.

The Vicious Circle

One can mention many other modes of land alienation. The above examples are given only to show the processes that lead to alienation of the people's livelihood. Because of the unequal nature of the encounter, also the reaction of these systems to the problems that the process causes differs. That too is based on their worldview on land and the natural resources all of whose dependants feel the negative impact of the transition from the traditional to the modern economy and new values.

But the rural poor, particularly the tribes and other forest dwellers feel its impact more than the remaining groups do because it is an attack on their tradition of judicious use of resources and on the systems they had developed to manage land, forests and CPRs as their renewable sustenance. Loss of their sustenance begins the vicious circle of impoverishment that forces the dependants of these resources to overexploit them and cause further environmental degradation and more poverty. As the former Brazilian President Fernando Henrique Cardoso (1998) said, the first danger to the environment from people's impoverishment is loss of biodiversity and linked to it, loss of the values through which the communities depending on it had managed the resource as renewable. Studies show that loss of this value system or ideology is basic to the vicious circle that leads to further environmental degradation.

But reaction to this process differs according to the class one belongs to and one's ideology or culture. To the urban middle class land alienation and environmental degradation are loss of their recreational spaces while to the rural, particularly tribal, communities it results in loss of their livelihood and consequent impoverishment from which follows further land alienation and destruction of more natural resources. Conflicts are a natural consequence of this contradiction. The first step of this process is impoverishment of the economic status they are reduced to by the alienation of their sustenance. It begins with landlessness. Then comes joblessness. For example, studies of families displaced by development projects show that in Andhra Pradesh in South India, the proportion of the landless rose from 10.9 per cent before the project to 36.5 per cent after it and in Assam in the Northeast from 15.56 to 24.38 per cent. Even among those who retain land, the average area owned declined, for example in Assam from 1.2 to 0.6 hectares per family. In every state most small and marginal farmers became landless, and medium farmers joined the ranks of small and marginal farmers. They also witnessed a decline in the support mechanisms such as the number of irrigation ponds and wells, poultry, cattle, and draught animals that used to supplement their agricultural income declined.

Joblessness is the next step. The land and other resources that are alienated from them used to provide them work. They lose this resource with no alternative to take its place. Joblessness resulting from it takes two forms. The first is lower access to work and the second is downward occupational mobility. In Andhra Pradesh, for example, 83.72 per cent of the land losers used to work on their land or elsewhere before its loss. After land loss access to work declined to 41.61 per cent. In West Bengal it declined from 91.02 to 53.18 percent and in Assam from 77.27 to 56.41 percent. The second is downward occupational mobility. In most states, more than 50 percent of the land losers who were cultivators before it became landless agricultural labourers or daily wage earners after land loss.

Also displacement can continue as a result of environmental degradation. For example, a new industry often forces people to move out of its neighbourood because after its construction environmental or other consequences such as fly ash and dust generated by the thermal, aluminium, nuclear, cement and other plants destroy the land around it and render it unusable. Its dependants cannot sustain themselves on it and are forced to move out. Also the noise and dust pollution and constant blasts in the coalmines often force people to leave their homes.

Absorbing a New Culture

The changes do not remain external but enter the community itself through the internalisation of the dominant culture. The major change is the culture the community in general and its elite in particular internalise, of viewing their sustenance as commodity alone. It is seen firstly in the demand the leaders make that individual ownership become the norm in their communities. For example, in the Garo tribe of East Garo in Meghalaya in Northeast India the leaders accepted the culture of individual ownership in the 1980s. A study two decades later shows that 30 percent of the tribal families in this district had become landless since their elite had monopolised much of their land. These changes also have gender implications. As stated above, even the matrilineal societies are patriarchal. Their leaders absorbs the culture of greater patriarchy and express it in their land relations. That can be seen among the Garo who are a matrilineal tribe but individual ownership is through men. Among the Khasi of Megalaya who too are a matrilineal tribe, the male leaders who control the village council exploit their power to their own advantage and turn community-owned land into their private property (Mukhim forthcoming). Such change of gender attitudes is seen in other tribes too in the manner in which men interpret their customary law and property relations in their own favour.

Communities thus deprived of their resources absorb the same culture in another form. The first is the vicious circle of viewing their resources as a commodity alone. Once they are deprived of their resource and are impoverished, for sheer survival they overexploit the same resource for an income. For example, studies in all the tribal areas show that once they lose their land, the deprived families fall back on their forests that they had preserved for centuries and cut trees for sale as firewood or timber, and cause more deforestation.

The second to view their own bodies as a commodity. For example, 49 percent of the families displaced by development projects in West Bengal and 56 percent in Assam pulled their children out of school in order to turn them into child labourers. Women began to view their bodies only as a source of income. Because of it prostitution grew enormously among the families that had lost their land. All these instances point to a major change in subaltern culture. These communities lose hope in their future and think only of the present. As a result, children who are an asset for the future become commodities only for the present and are used as a source of income for survival. The same is the view of women's bodies. In other words, women and children become commodities more than men do.

This chapter which is an overview of the changes in land relations, has shown the new culture that grown as their result. It shows that imposition of another culture on a traditional group can result in a culture that is destructive of a community in general and of women in particular. The solution is not either going back to their tradition by opposing modernisation or absolutising the modern system. One cannot prevent all individual ownership either. One has to find an alternative in beginning with the tradition community values and combining them with the traditional community. Tradition has to be modernised and not replaced completely.

STRENGTHEN ADAPTIVE CAPACITY OF FORESTS

Negatively affect forests and many of their plant and animal species. In addition, they may negatively affect the availability of other resources, necessary for species survival. Current forest composition and structure are however, the result of past changes in climate and shows that forests and their species have an inherent capacity to adapt to change. The main differences of current climate change with historic changes are the increased rate of these changes and the degraded and fragmented state of the remaining forests, which reduces the capacity of the species and ecosystems to adapt. The challenge is to help species and ecosystems to adapt to climate change while at the same time ensuring that ecosystem services are maintained. This will require the identification of the changes to which the forest will need to adapt.

Locally, changes may be disastrous, unless climate, ecosystem and species changes are accompanied by adjustments in the local social and economic systems. For example, increased occurrence of severe fires will require greater collective action to prevent fires as well as improved weather and fire danger forecast services. Companies producing furniture of high value species from natural forests, whose natural regeneration under changed climate conditions has become increasingly difficult, may have to change geographic range for their inputs, or change to other species and/or other processing procedures. Communities and private landowners depending on local forests may have to change livelihoods after severe hurricane damage.

Nationally or at the landscape level, changes may be slower and less disastrous in the short term. New challenges include the identification of those species groups and ecological processes that are essential for the most important ecosystem services. This would include in most cases identification of water catchment areas (hydrogeology) and the role of forests in maintaining water quality and quantity. It will be important to increase the probability that changing ecosystems will continue to provide the important services and goods. In particular, ecosystems in geographic locations at the extreme limits of climatically well-defined areas, such as mountainous forests, rangelands and boreal forests, are likely to be severely affected and may disappear. Some authors suggest that maintaining functional diversity and 6 composition will preserve ecosystem services, while others found that different functional groups will react differently to environmental changes, indicating that climate change may favour some functional groups over others. More research is needed however, to identify those functional groups essential for the desired ecosystem services and goods in particular areas and to understand how these can be conserved and protected.

Reduce Risk and Intensity of Pest, Disease and Fire Outbreaks

Reducing the climate induced risk of pests, diseases and fire outbreaks, in particular, in dry areas and less diverse forests will be a major environmental challenge. Breeding of more resistant or more resilient varieties is a medium to long-term solution for plantation species, although, that introduces new risks because strengthening the adaptive capacity of a species for one trait may weaken it to other traits. Identifying species for their "realized fitness" - for example, varieties of a species that survived insect attacks, diseases or fires, similar to the expected events in a particular region - and then facilitating their migration to the area of interest, may be another strategy. In both cases, identification of the traits that will increase resistance or resilience will be important as will be replicating those traits over generations and successfully introducing the species or varieties in the area of interest, without introducing new problems (such as undesired invasion).

Predicting future changes in pest and disease outbreaks and adjusting management accordingly is another option, which requires the development and validation of models that reliably predict impacts under different climate and management scenarios. A further option is the identification and implementation of forest management systems that are known or thought to reduce the risks of pests, diseases and/or fires.

While there are several well known means to protect forests and plantations, in many cases these are not applied for a variety of reasons, or are not applied to those forests most in need. The challenges are to identify and address the reasons for the lack of application of management techniques and to adjust management options to the threats in a participatory, socially and economically acceptable manner.

CHANGES IN SOCIOECONOMIC ENVIRONMENT

Risk of Migration into Forest Areas

Climate change will affect all people but in particular, rural people that depend on nature for their livelihoods, and poverty stricken communities in the urban-rural interface that are often subjected to the consequences of extreme weather events. Climate change is expected to change the aptitude of lands for specific crops, cause problems of droughts, fire and flooding and may drive many people from their lands. These people are likely to either go to cities to look for jobs, often adding to urban poverty, or to other rural areas to look for other lands where they may be able to continue their agricultural livelihoods or find employment in the agricultural sector.

The surge of interest in fuels from biomass (*e.g.* corn, sugarcane and oil palm) adds another dimension to this migration. The purchasing of land, often based on speculation, in the hope of selling later for higher prices to investors interested in biofuel production, may cause migration. The expected high incomes from biofuels may also motivate landowners to convert their forests into energy plantations, oftentimes in an unsustainable manner. On the other hand, if well planned, biofuels could also help avoid or reduce migration by providing off-farm employment.

Forest use values, even in the case of the most successful enterprises, will not be able to compete with oil palm or other energy crops in those lands suitable for the crops. Legal definition of user and owner rights of forest areas and the mechanisms to defend those rights will be important elements of strategies to prevent unauthorized entrance into forests. Market mechanisms that restrict trade of products from companies that do not show social and environmental responsibility in their production and purchase policies may be another strategy.

An individual forest user or owner will find it difficult to influence legislation, their implementation or the way that markets function. Collaboration with other stakeholders, neighbours, value chain members, and state administrators will be essential to the development of adequate measures to reduce the conversion and degradation of forests. Forest users and owners, however, have a longstanding tradition of independence and in the past have not shown tendencies to such collaboration.

Lack of trust (often justified), has often hampered relations between different stakeholders in the forest and environmental sectors. Building sufficient trust to facilitate collaboration may be the biggest challenge of all for future forest management and needs the collaboration of all actors involved.

ECOSYSTEM OF TROPICAL RAINFOREST

A tropical rainforest is an ecosystem type that occurs roughly within the latitudes 28 degrees north or south of the equator (in the equatorial zone between the Tropic of Cancer and Tropic of Capricorn). This ecosystem experiences high average temperatures and a significant amount of rainfall. Rainforests can be found in Asia, Australia, Africa, South America, Central America, Mexico and on many of the Pacific, Caribbean, and Indian Ocean islands. Within the World Wildlife Fund's biome classification, tropical rainforests are thought to be a type of tropical wet forest (or tropical moist broadleaf forest) and may also be referred to as *lowland equatorial evergreen rainforest*.Tropical rainforests can be characterized in two words: warm and wet. Mean monthly temperatures exceed 18 °C (64 °F) during all months of the year. Average annual rainfall is no less than 168 cm (66 in) and can exceed 1,000 cm (390 in) although it typically lies between 175 cm (69 in) and 200 cm (79 in). This high level of precipitation often results in poor soils due to leaching of soluble nutrients.

Tropical rainforests are unique in the high levels of biodiversity they exhibit. Around 40 per cent to 75 per cent of all biotic species are indigenousto the rainforests. Rainforests are home to half of all the living animal and plant species on the planet. Two-thirds of all flowering plants can be found in rainforests. A single hectare of rainforest may contain 42,000 different species of insect, up to 807 trees of 313 species and 1,500 species of higher plants.

Tropical rainforests have been called the "jewels of the Earth" and the "world's largest pharmacy", because over one quarter of natural medicines have been discovered within them. It is likely that there may be many millions of species of plants, insects andmicroorganisms still undiscovered in tropical rainforests.

Tropical rainforests are among the most threatened ecosystems globally due to large-scale fragmentation due to human activity. Habitat fragmentation caused by geological processes such as volcanism and climate change occurred in the past, and have been identified as important drivers of speciation.

However, fast human driven habitat destruction is suspected to be one of the major causes of species extinction. Tropical rain forests have been subjected to heavy logging and agricultural clearance throughout the 20th century, and the area covered by rainforests around the world is rapidly shrinking.

HISTORY

Tropical rainforests have existed on Earth for hundreds of millions of years. Most tropical rainforests today are on fragments of the Mesozoic era supercontinent of Gondwana. The separation of the landmass resulted in a great loss of amphibian diversity while at the same time the drier climate spurred the diversification of reptiles. The division left tropical rainforests located in five major regions of the world: tropical America, Africa, Southeast Asia, Madagascar, and New Guinea, with smaller outliers in Australia. However, the specifics of the origin of rainforests remain uncertain due to an incomplete fossil record.

TYPES OF TROPICAL RAINFOREST

Several types of forest comprise the general tropical rainforest biome:

- Lowland equatorial evergreen rain forests are forests which receive high rainfall (more than 2000 mm, or 80 inches, annually) throughout the year. These forests occur in a belt around the equator, with the largest areas in the Amazon Basin of South America, the Congo Basin of Central Africa, Indonesia, and New Guinea.
- Moist deciduous and semi-evergreen seasonal forests, receive high overall rainfall with a warm summer wet season and a cooler winter dry season. Some trees in these forests drop some or all of their leaves during the winter dry season. These forests are found in parts of South America, in Central America and around the Caribbean, in coastal West Africa, parts of the Indian subcontinent, and across much of Indochina.
- Montane rain forests, some of which are known as cloud forests, are found in cooler-climate mountain areas. Depending on latitude, the lower limit of montane rainforests on large mountains is generally between 1500 and 2500 m while the upper limit is usually from 2400 to 3300 m.

- Flooded forests, seven types of flooded forest are recognized for Tambopata Reserve in Amazonian Peru:
- *Permanently waterlogged swamp forest*—Former oxbow lakes still flooded but covered in forest.
- *Seasonally waterlogged swamp forest*—Oxbow lakes in the process of filling in.
- *Lower floodplain forest*—Lowest floodplain locations with a recognizable forest.
- *Middle floodplain forest*—Tall forest, flooded occasionally.
- *Upper floodplain forest*—Tall forest, rarely flooded.
- *Old floodplain forest*—Subjected to flooding within the last two hundred years.
- *Previous floodplain*—Now *terra firme*, but historically ancient floodplain of Tambopata River.

FOREST STRUCTURE

Rainforests are divided into different strata, or layers, with vegetation organized into a vertical pattern from the top of the soil to the canopy Each layer is a unique biotic community containing different plants and animals adapted for life in that particular strata. Only the emergent layer is unique to tropical rainforests, while the others are also found in temperate rainforests.

Forest Floor

The forest floor, the bottom-most layer, receives only 2 per cent of the sunlight. Only plants adapted to low light can grow in this region. Away from riverbanks, swamps and clearings, where dense undergrowth is found, the forest floor is relatively clear of vegetation because of the low sunlight penetration. This more open quality permits the easy movement of larger animals such as: ungulates like the okapi (*Okapia johnstoni*), tapir (*Tapirus* sp.), Sumatran rhinoceros (*Dicerorhinus sumatrensis*), and apes like the western lowland gorilla (*Gorilla gorilla*), as well as many species of reptiles, amphibians, and insects. The understory also contains decaying plant and animal matter, which disappears quickly, because the warm, humid conditions promote rapid decay. Many forms of fungi growing here help decay the animal and plant waste.

Understory Layer

The understory layer lies between the canopy and the forest floor. The understory is home to a number of birds, small mammals, insects, reptiles, and predators. Examples include leopard(*Panthera pardus*), poison dart frogs (*Dendrobates* sp.), ring-tailed coati (*Nasua nasua*), boa constrictor (*Boa constrictor*), and many species of Coleoptera. The vegetation at this layer generally consists of shade-tolerant shrubs, herbs, small trees, and large woody vines which climb into the trees to capture sunlight. Only about 5 per cent of sunlight breaches the canopy to arrive at the understory causing true understory plants to seldom grow to 3 m (10 feet). As an adaptation to these low light levels, understory plants have often evolved much larger leaves. Many seedlings that will grow to the canopy level are in the understory.

Canopy Layer

The canopy is the primary layer of the forest forming a roof over the two remaining layers. It contains the majority of the largest trees, typically 30–45 m in height. Tall, broad-leaved evergreen trees are the dominant plants. The densest areas of biodiversity are found in the forest canopy, as it often supports a rich flora of epiphytes, including orchids, bromeliads, mosses and lichens.

These epiphytic plants attach to trunks and branches and obtain water and minerals from rain and debris that collects on the supporting plants. The fauna is similar to that found in the emergent layer, but more diverse. It is suggested that the total arthropod species richness of the tropical canopy might be as high as 20 million. Other species habituating this layer include many avian species such as the yellow-casqued wattled hornbill (*Ceratogymna elata*), collared sunbird (*Anthreptes collaris*), African gray parrot(*Psitacus erithacus*), keel-billed toucan (*Ramphastos sulfuratus*), scarlet macaw (*Ara macao*) as well as other animals like the spider monkey (*Ateles*sp.), African giant swallowtail (*Papilio antimachus*), three-toed sloth (*Bradypus tridactylus*), kinkajou (*Potos flavus*), and tamandua (*Tamandua tetradactyla*).

Emergent Layer

The emergent layer contains a small number of very large trees, called emergents, which grow above the general canopy, reaching heights of 45–55 m, although on occasion a few species will grow to 70–80 m tall. Some examples of emergents include: *Balizia elegans*, *Dipteryx panamensis*, *Hieronyma alchorneoides*, *Hymenolobium mesoamericanum*, *Lecythis ampla* and*Terminalia oblonga*. These trees need to be able to withstand the hot temperatures and strong winds that occur above the canopy in some areas. Several unique faunal species inhabit this layer such as the crowned eagle (*Stephanoaetus coronatus*), the king colobus (*Colobus polykomos*), and the large flying fox (*Pteropus vampyrus*). However, stratification is not always clear. Rainforests are dynamic and many changes affect the structure of the forest. Emergent or canopy trees collapse, for example, causing gaps to form. Openings in the forest canopy are widely recognized as important for the establishment and growth of rainforest trees. It's estimated that perhaps 75 per cent of the tree species at La Selva Biological Station, Costa Rica are dependent on canopy opening for seed germination or for growth beyond sapling size, for example.

ECOLOGY AND CLIMATE

Most tropical rainforests are located around and near the equator, therefore having what is called an equatorial climate characterized by three major climatic parameters: temperature, rainfall, and dry season intensity Other parameters that affect tropical rainforests are carbon dioxide concentrations, solar radiation, and nitrogen availability. In general, climatic patterns consist of warm temperatures and high annual rainfall. However, the abundance of rainfall changes throughout the year creating distinct wet and dry seasons. Rainforests are classified by the amount of rainfall received each year, which has allowed ecologists to define differences in these forests that look so similar in structure.

According to Holdridge's classification of tropical ecosystems, true tropical rainforests have an annual rainfall greater than 800 cm and annual temperature greater than 24 degrees Celsius. However, most lowland tropical rainforests can be classified as tropical moist or wet forests, which differ in regards to rainfall. Tropical rainforest ecology- dynamics, composition, and function- are sensitive to changes in climate especially changes in rainfall. The climate of these forests is controlled by a band of clouds called the Intertropical Convergence Zone located near the equator and created by the convergence of the trade winds from the northern and southern hemispheres.

The position of the band varies seasonally, moving north in the northern summer and south in the northern winter, and ultimately controlling the wet and dry seasons in the tropics. These regions have experienced strong warming at a mean rate of 0.26 degrees Celsius per decade which coincides with a global rise in temperature resulting from the anthropogenic inputs of greenhouse gases into the atmosphere. Studies have also found that precipitation has declined and tropical Asia has experienced an increase in dry season intensity whereas Amazonia has no significant pattern change in precipitation or dry season. Additionally, El Niño-Southern Oscillation events drive the interannual climatic variability in temperature and precipitation and result in drought and increased intensity of the dry season.

As anthropogenic warming increases the intensity and frequency of ENSO will increase, leaving tropical rainforest regions susceptible to stress and increased mortality of trees.

SOIL TYPES

Soil types are highly variable in the tropics and are the result of a combination of several variables such as climate, vegetation, topographic position, parent material, and soil age Most tropical soils are characterized by significant leaching and poor nutrients; however there are some areas that contain fertile soils. Soils throughout the tropical rainforests fall into two classifications which include the ultisols and oxisols.

Ultisols are known as well weathered, acidic red clay soils, deficient in major nutrients such as calcium and potassium. Similarly, oxisols are acidic, old, typically reddish, highly weathered and leached, however are well drained compared to ultisols. The clay content of ultisols is high, making it difficult for water to penetrate and flow through. The reddish colour of both soils is the result of heavy heat and moisture forming oxides of iron and aluminum, which are insoluble in water and not taken up readily by plants.

Soil chemical and physical characteristics are strongly related to above ground productivity and forest structure and dynamics. The physical properties of soil control the tree turnover rates whereas chemical properties such as available nitrogen and phosphorus control forest growth rates.

The soils of the eastern and central Amazon as well as the Southeast Asian Rainforest are old and mineral poor whereas the soils of the western Amazon (Ecuador and Peru) and volcanic areas of Costa Rica are young and mineral rich. Primary productivity or wood production is highest in western Amazon and lowest in eastern Amazon which contains heavily weathered soils classified as oxisols. Additionally, Amazonian soils greatly weathered, making them devoid of minerals like phosphorus, potassium, calcium, and magnesium, which come from rock sources.

However, not all tropical rainforests occur on nutrient poor soils, but on nutrient rich floodplains and volcanic soils located in the Andean foothills, and volcanic areas of Southeast Asia, Africa, and Central America. Oxisols, infertile, deeply weathered and severely leached, have developed on the ancient Gondwanan shields. Rapid bacterial decay prevents the accumulation of humus. The concentration of iron and aluminum oxides by the laterization pro cess gives the oxisols a bright red colour and sometimes produces minable deposits. On younger substrates, especially of volcanic origin, tropical soils may be quite fertile.

Nutrient Recycling

This high rate of decomposition is the result of phosphorus levels in the soils, precipitation, high temperatures and the extensive microorganism communities. In addition to the bacteria and other microorganisms, there are an abundance of other decomposers such as fungi and termites that aid in the process as well. Nutrient recycling is important because below ground resource availability controls the above ground biomass and community structure of tropical rainforests.

These soils are typically phosphorus limited, which inhibits net primary productivity or the uptake of carbon. The soil contains tiny microbial organisms such as bacteria, which break down leaf litter and other organic matter into inorganic forms of carbon usable by plants through a process called decomposition. During the decomposition process the microbial community is respiring, taking up oxygen and releasing carbon dioxide. The decomposition rate can be evaluated by measuring the uptake of oxygen. High temperatures and precipitation increase decomposition rate, which allows plant litter to rapidly decay in tropical regions, releasing nutrients that are immediately taken up by plants through surface or ground waters. The seasonal patterns in respiration are controlled by leaf litter fall and precipitation, the driving force moving the decomposable carbon from the litter to the soil. Respiration rates are highest early in the wet season because the recent dry season results in a large percentage of leaf litter and thus a higher percentage of organic matter being leached into the soil.

Buttress Roots

A common feature of many tropical rainforests is the distinct buttress roots of trees. Instead of penetrating to deeper soil layers, buttress roots create a wide spread root network at the surface for more efficient uptake of nutrients in a very nutrient poor and competitive environment.

Most of the nutrients within the soil of a tropical rainforest occur near the surface because of the rapid turnover time and decomposition of organisms and leaves. Because of this, the buttress roots occur at the surface so the trees can maximize uptake and actively compete with the rapid uptake of other trees. These roots also aid in water uptake and storage, increase surface area for gas exchange, and collect leaf litter for added nutrition. Additionally, these roots reduce soil erosion and maximize nutrient acquisition during heavy rains by diverting nutrient rich water flowing down the trunk into several smaller flows while also acting as a barrier to ground flow. Also, the large surface areas these roots create provide support and stability to rainforests trees, which commonly grow to significant heights. This added stability allows these trees to withstand the impacts of severe storms, thus reducing the occurrence of fallen trees.

FOREST SUCCESSION

Succession is an ecological process that changes the biotic community structure over time towards a more stable, diverse community structure after an initial disturbance to the community. The initial disturbance is often a natural phenomenon or human caused event. Natural disturbances include hurricanes, volcanic eruptions, river movements or an event as small as a fallen tree that creates gaps in the forest. In tropical rainforests, these same natural disturbances have been well documented in the fossil record, and are credited with encouraging speciation and endemism.

2

Sustainable Ecosystems in Forest

The desired outcome of the Forest Service's activities is ecosystem sustainability. Sustainability should be viewed more as a journey than a fixed destination. In this context, sustainable forest management has three component parts: social, economic, and ecological. These three components are interrelated and must be considered together when making assessments of sustainable resource management.

THE LEGAL FOUNDATION OF THE FOREST SERVICE

The 1897 Organic Act's statement of purpose embedded the principle of sustainability. In part, the national forests were established to "...improve and protect the forest within the boundaries, or for the purpose of securing favorable conditions of water flows, and to furnish a continuous supply of timber for the use and necessities of citizens of the United States...". Timber was allowed to be sold "For the purpose of preserving the living and growing timber and promoting the younger growth on national forests...". While these notions and principles of sustainability were not exhaustive in their meaning, the Organic Act reflects early understanding that social health and public welfare are affected by and dependent upon natural resources and the management of the landscapes in which they occur.

PROMOTING ECOSYSTEM SUSTAINABILITY

Other foundational Federal environmental laws, together with an eclectic set of environmental state laws of our 50 states, have further embedded the principles of sustainable resource management. Myriads of local and county ordinances designed to protect the environment concurrent to rural and urban development also contribute to the U.S. institutional capacity to regulate development through threshold standards based on resource sustainability.

Collectively, these laws cover a range of values that characterize the ecosystems people occupy and that provide for human needs. Some legislation is agency specific (such as the National Forest Management Act which applies to the F.S.), some unique to states (such as the State of Oregon Forest

Practices); and some national in scope but applies differently to ownerships (such as the T&E Species Act). Nevertheless, together, this body of environmental law not only contains minimal thresholds and standards, but also provides for both direction and enablement of activities designed to protect the environment concurrent to meeting societal needs. The recent emergence of certification schemes and adoption of national criteria and indicators of sfm, are reflective of continued adaptive management designed to protect the environment concurrent to development.

FORESTED LAND IS JURISDICTIONALLY FRAGMENTED

Context is more important than content. Understanding the context in which US forestry related work is accomplished is important. No matter how sustainability is defined, it remains a subset of a larger reality that cannot be independently achieved. U.S. forested ecosystems are managed by a decentralized group of Federal, state, industrial, and private entities which is a challenge for sustainable forest management. Managers need to know how their activities affect the sustainability of shared values that cross their respective ownerships.

States Independently Developed Forest Measurement Systems

The 737 million acres of forest land covers the jurisdictions of our 50 states. States have a long tradition of establishing government entities (agencies, bureaus, offices, departments, commissions, etc) through which public programmes are focused on forest resources. For example in 2000, states had created 204 governing or advisory bodies, variously labeled as boards, councils, committees, or commissions. States independently developed environmental data standards and protocols, and they are not compatible or comparable.

Land Owners Independently Developed Forest Measurement Systems

Land ownership patterns within states add to the operational complexity of the U.S. forest working environment. For example, of the 489 million acres of commercial forest land, 58% is managed by over six million land owners; 15% by forest industries; 17% by USFS; and 10% by other public agencies. In 1999, nearly 400 public and private entities at the state level were engaged in gathering, managing or distributing information about forests (National Association of State Foresters, 1999). These ownerships also independently developed forest measurement systems that are largely incompatible with each other.

Agencies Themselves Often Have Measurement Systems

As occurs at the state level, the organizational landscape of the Federal government is also composed of many divisions. For example, the US Forest

Service manages 17% of the nations commercial forest land. The Forest Service headquarters located in Washington, DC, administers these lands through nine Regional Offices. The nine Regions are administered through 125 National Forest Offices. The 125 National Forests are administered through 575 District Offices. The Forest Service has a national grid point inventory system for all U.S. forested land. However, measurement indices lack social, economic, and indices for non-commodity values. While some corporate data exists and progress is being made in developing more corporate data, at the Forest level, data standards and protocols were independently developed, resulting in the development of incomparable or incompatible data. This situation typifies most land management agencies. Added to this operational complexity are the different missions, objectives, and regulatory frameworks of respective forest management organizations.

Related to movement towards ecosystem sustainability, the value of our nation's resource management organizations is not to be found in our management diversity, mission, or even our legal framework - however important: but in the measurable outcomes that reflect our collective actions.

HEALTHY ECOSYSTEMS AND SUSTAINABLE ECONOMIES

Forest land trends have been stable. The current U.S. forest land area of 747 million acres amounts to about 71% of the area that was forested at the time of European settlement in 1630. Since 1630, about 297 million acres of forest land have been converted to other uses—mainly to agriculture and primarily in the North and South. After 1910 agricultural clearing subsided and the amount of forest area stabilized. The trend shows that the South and pacific-West are experiencing modest declines in forest land, as population shifts to those areas while forest area in the North is increasing.

Growing stock and timber harvest trends have been stable. During the last 10 years timber harvest has not varied much. However, harvest from Forest Service lands has decreased (2,210 to 829 mcuft), while harvest from non-industrial lands has increased (8,124 to 9,515 mcuft). Shifts in reduced harvest from Forest Service lands reflect shifts in societal values promoting recreation, tourism, and old-growth habitat for threatened and endangered species. Many of the problems we face in our national forests today defy simple administrative solutions. One serious problem is the health of our forest ecosystems. Many national forest lands have a moderate to severe risk of catastrophic fire. Forest health speaks to the importance and need for cross-jurisdictional planning designed to sustain US forested ecosystems and to address common concerns.

Along Administrative Lines

Although we are making progress in managing landscapes more holistically, our collective approach has often fallen short of maintaining the health and sustainability of landscapes (Interagency Ecosystem Management Task Force, 1995). Often times, neighboring managers within a landscape do not know how their actions affect landscape sustainability. This confusion undermines collaborative processes, causes lack of trust, litigation, and often postures managers in a crisis management mode. Several things mitigate against collaborative processes:

- Communication: We lack common data standards and collection protocols between neighboring land managers.
- Consensus: There is little public consensus on how to frame public dialogue regarding sfm
- Analysis: We are weak in integrating environmental, economic and social issues at multiple scales.
- Harmonizing plans: We have difficulty in developing common goals and shared responsibilities with our neighbors.
- Funds: We have difficulty focusing scarce funds to highest priority areas.

FOREST SERVICE PLANNING PROCESSES ARE UNDERGOING CHANGE

The National Forest Management Act that guides FS planning processes is currently under revision. The proposed rule stresses the importance of ecological principles in sound, multiple-use, sustainable forest management. It presents a system designed to better incorporate science, research, and interested members of the public into national forest management. The proposed rule also requires that forest and grassland plans be consistent to national strategic documents; to set goals for management; and to establish programmatic standards to follow in pursuit of strategic direction. A final decision on the rule is expected sometime this year.

National Forest System Planning

The proposed revisions to the FS planning process express the Agency's commitment to the principles of sfm. The proposed rule "...makes sustainability the foundation for National Forest System planning and management and establishes requirements for implementation, monitoring, evaluation, amendment, and revision of land and resource management plans." The first priority for planning will be to maintain and restore the sustainability of watersheds, forests, and rangelands for present and future generations.

Central Reference Points for Planning

Companion to the notion of sustainability, the proposed rule requires that "desired future conditions" and outcomes associated with them be used as the central reference points for planning. Managers would establish pathways to the desired future conditions and orient performance measures, monitoring, and budgeting to those pathways. Land and resource management plan direction would seek to prevent the need for a species being listed under the T&ES Act.

A Spatial Approach

The proposed rule fosters a spatial approach to planning and analysis-strengthening linkages across ownership and between scales. Both large-scale and/or small-scale assessments are to be used as the means of obtaining necessary information. Planning is to emphasize ecological boundaries for assessment and consider their social meaning. Managers will address all Federal lands within the area and work with affected agencies and move toward integrated administration of jurisdictionally fragmented areas. These steps are designed to encourage managers to harmonize respective resource management plans through landscape planning independent of jurisdictional lines, thus facilitating collaborative stewardship.

Planning Processes Encourage Shared Decision-Making

Collaborative stewardship efforts will prove to be the vehicles of choice for managers to minimize costly litigation; to reconnect and heal impaired landscapes; and to foster community understanding and the development of shared goals. The proposed rule fosters opportunities to move away from a Federally focused governmental decision-making structure towards a collaborative design that shares responsibility with others. Managers are to provide opportunities for continuous, early, open, and frequent involvement to develop a shared vision and achieve an understanding of proposed actions and their expected outcomes.

The responsible official is to foster relationships and create partnerships. The FS official becomes a convener, facilitator, leader or participant, in collaborative processes. The desired outcome of collaborative stewardship is healthy ecosystems and sustainable economies. Good collaboration is where people work together better such that the outcomes of collective stewardship efforts improve and move us toward the national goal of ecosystem sustainability. Key to success is working with willing partners, respecting each other's unique roles, responsibilities, land use objectives, and limitations of the land.

Planning Processes

The proposed rule strengthens the role of science in the planning process. Scientists will assist the responsible official in understanding and applying relevant information. Scientists will be involved in issue identification; in helping to design and evaluate monitoring and inventory strategies and protocols; and in reviewing planning processes to ensure consistency in the application and interpretation of best available information. Having a planning system that: is based on sustainability; that uses desired future conditions as a reference point; that uses broad-scale assessments; encourages collaborative processes; and that incorporates science into decision making, has great potential to strengthen the ecosystem management approach, which is the FS means along the road towards sustainability. However, without a common language of data standards and protocols, cross-jurisdictional communication issues will remain problematic.

Collaborative Processes

Most entities that manage forests have data but the data is largely incomparable or incompatible. Currently the US annually spends 100's of millions of dollars on data collection without the capability to assess the sustainability of our forested ecosystems, or to effectively and efficiently conduct collaborative processes designed to address common concerns and objectives. We lack a unifying measurement framework that crosses ownerships and that integrates social, economic, and ecological indices of sustainability suitable for collaborative processes.

The Montreal Process Criteria And Indicators Provides a Unifying Framework

The criteria and indicators of sustainable forest management, internationally agreed to by the 12 countries of the Montreal Process are providing a unifying national framework for public dialogue (Criteria and Indicators, 1995). It provides the means for measuring sustainable forest management "vital signs" across a diverse landscape.. The FS has adopted the Montreal Process C&I and is leading a national, interagency/ngo Roundtable initiative to implement them. The Roundtable consists of 32 Federal and state agencies that affirmed their commitment to sfm. These partners recognize that sfm (at a national scale) is beyond the capability of any one agency and must be accomplished through partnerships. The C&I are a relatively simple statement of seven key goals (criteria), with associated measurements (indicators), designed to promote an understanding of what constitutes sustainable management of temperate and boreal forests.

Measurement Systems Need To Be Simplified

The Montreal Process (MP) framework of measures provides focal point to develop common data standards and prototypes for the indicators. The Roundtable partners are working with the Federal Geographic Committee to establish data standards and protocols, interagency budget submissions to populate the MP framework with data, and to help establish a central data repository facilitating the sharing of information. In effect, using the MP C&I

measurement framework is simplifying our measurement systems and will establish a baseline of compatible and comparable data.

A Common Language Of Measurements For Partners

The Montreal Process C&I enjoy broad-based support from a diverse group of national stakeholders. A desired outcome of the Roundtable process is to establish the MP measures as national standards and to embed measures in agency programmes.

This will help promote communication as well as help frame public dialog on sfm and movement towards consensus. The MP C&I strengthen shared analysis through the integration of environmental, economic and social issues at multiple scales.

The MP C&I enable partners to see how they contribute to sfm enabling the development of common goals and shared responsibilities with respective neighbors. At the landscape, regional, or country level, sustainable forest management can only be achieved through the collective result of actions on all lands within the respective area of interest as measured by criteria and indicators. Any particular management regime's sustainability is best judged by the *quality of management* versus how it responds to all 67 of the MP indicators. Finally, the MP C&I help to focus scarce funds to highest priority areas and provide a framework for interagency budget proposals.

Link Interrelated Management Processes

The MP C&I measures provide a unifying framework for strengthening the linkages and relationships between parallel and interrelated processes designed to meet societal needs concurrent to protecting ecosystem sustainability. The Forest Service (FS) model flow chart shows key linkages between processes for decision making.

NATIONAL LEVEL ASSESSMENTS PROVIDE CONTEXT FOR SUB-NATIONAL LEVEL WORK

Effective higher-level assessments provide context for work at sub-national scales and focus on indices of sustainability. The FS conducts inventories and completes assessments based on criteria and indicators. Inventories and assessments are done both at the national and sub-national scales. A good national assessment will provide the context for landscape planning, and sub-national broad-scale assessments provide the context for project level work. Sustainability is the desired outcome of FS activities; therefore key indices of sustainability focus inventory and assessment work designed to address multiple business management practices associated with sustainable resource management. Information in both national and broad-scale assessments will be organized using the seven MP Criteria as an organizing framework.

Sub-National Project Plans

Based on national assessments, the FS develops a strategic plan that disclosed to the Congress and American people what the FS proposes to do for appropriated funds received. The goals of the strategic plan have been cross-walked to the seven MP Criteria. The strategic plan includes specific performance measures to measure progress and enable accountability. The work identified in the strategic plan is carried out in the context of sub-national Forest and Grassland Management Plans that detail environmental impacts and that are supported by more site specific project level plans. Project level plans are designed to carry out specified work associated with accomplishment of annual goals.

Annual Budget Allocations

The Congress allocates the FS an annual budget based on the strategic plan that discloses desired future conditions. The FS disaggregates this funding to field offices. For example, the WO funds the Regions who fund the Forests, who fund the Districts, who execute project level work in local communities. Project level work is designed to contribute to accomplishment of desired future conditions at both the national and sub-national levels.

LOCAL ACTIVITIES REFLECTS NATIONAL DIRECTION

Monitoring and inventory work can become a black hole where enough is never enough and more is always possible. Sustainability is the desired outcome of Agency activities, thus performance measures associated with desired future conditions should be linked to vital sign indices of sustainable forest management. Local measures that are rooted in national standards enable value added understanding of information between scales. The formats for both the annual Forest and Regional Monitoring Reports will use the seven MP Criteria as a way to organize information. This will enable managers to see how their forests contribute to sustainability at higher scales.

Effective Monitoring and Management

The FS along with their partners are working to embed the MP C&I into respective national inventory systems. Partners are seeking to establish a national inventory platform on which respective partners contribute to gathering of sfm data appropriate to respective unique roles and responsibilities. Not all the MP indicators are applicable at the local scale. The FS has undertaken research to identify local level indicators of sustainability. These local measures will be organized under the seven MP criteria and be linked to national indicators as appropriate.

Much more can be said about the management model and the benefits of using the MP C&I as an organizing framework. Important is that the FS is not solely 'top down' or solely 'bottom up' management. Synergism occurs at all levels all the time feeding into adaptive management designed to meet emerging needs in carrying out the FS mission of serving people and caring for the land.

LOCAL CRITERIA AND INDICATOR MEASUREMENTS

Conservation and Maintenance of Soil and Water Resources. The national long-term objective is to improve and protect watershed conditions (U.S.D.A. Forest Service Strategic Plan - 2000 Revision). The national outcome performance measure is the trend in watershed meeting state water quality standards. Connecting local activities and performance to strategic goals helps to focus scarce resources to highest priority areas and enables accountability. Not everything needs to be 'hard wired', but clear linkages should be established. In this example, sub-national goals are expressed as achieving proper function (pfc) condition on all class three streams and above. Local measures include miles of stream that meet pfc.

Sustainability has been rightly critized as being a 'fuzzy idea'. The C&I simply identify key attributes that reflect societal values and science information. Additionally, sustainability cannot be assessed on the basis of only one C&I; rather, they need to be used together along with other considerations as appropriate. Existing conditions, benchmark goals, and related activities are noted and serve managers as a reference point from which to gauge progress.

Monitoring Trends Often Precipitate Legislation

The US approach to ecosystem sustainability is characterized by an eclectic set of national and state policies and Acts that have as presupposition sustainability. While these are non-regulatory in nature, they are usually accompanied with a complementary set of land use guidance and/or a system of best management practices. Changes in best management practices, regulation, and or the application of certification efforts are activities designed to achieve a desired future condition.

The major private sector organizations currently engaged in creating and implementing sustainable forestry standards in the United States include the American Forest and Paper Association, the American Tree Farm System, the National Woodland Owners Association, Rainforest Alliance's SmartWood programme, and Scientific Certification Systems. The American Forest & Paper Association's Sustainable Forestry Initiative details a set of principles and guidelines that are a requisite for membership in the Association. The SFI principles and guidelines are a mixture of general concepts and tangible measures of company performance in meeting broad SFI objectives. The presence of these programmes is in response to consumer demands for responsible land management. While there are differences, the private sector approach to sfm is compatible to the larger effort MP C&I initiative they are involved in.

Montreal Process Criteria and Indicators

Meeting increasing consumer demands from a fixed land and resource base will continue to precipitate the need for new approaches and solutions to emerging problems yet to be defined. It is likely that the success of the US FS and other agencies, will largely be decided through our ability to conduct an open, inclusive, and empowered planning process in which people (inclusive of our land managing neighbors) feel that they have the ability to influence the outcome through their involvement. Challenges include:

- To fashion a decision-making structure that is collaborative in design enabling the sharing of responsibility among all levels of government, interest groups, and neighboring land management entities, and that respects differing roles and responsibilities, land use objectives, and the limitations of the land
- To get all partners to commit to using the MP C&I and aggressively embed them in respective operational programmes.
- To develop agreed upon comparable and compatible data standards and protocols for the indicators.

FOREST ECOSYSTEM SERVICES BY LOCAL PEOPLE

Climate change is expected to increase the frequency and intensity of extreme weather events, such as hurricanes, torrential rains and droughts. Rural people often depend on emergency supplies during or just after such events. Forests, in many cases in the past, have provided such emergency supplies or safety nets *e.g.* wood for construction and repair of houses, woodfuel for cooking and fruits and other food to replace the lost crops. The need for these safety nets will further increase when climate change increases the loss of crops. Indigenous groups are often vulnerable to extreme events, especially those events that restrict access to the outside world and markets. However, in such cases, they can usually find sufficient emergency supplies from within the forest until access is restored. In addition, more people have become aware of the different ecosystem services and want to use such services even under non-extreme weather conditions. Forests as regulators of water quality and quantity have become ever more important, in particular, in areas with frequent droughts and/or frequent torrential rains that may cause erosion, sedimentation and flooding. In Central America, this function may be one of the main reasons for forest protection or restoration by private landowners even though it is possibly based on an erroneous perception of the benefits of the forest, since such functions may not be beneficial in some climate and soil conditions.

The impact of climate change on this ecosystem service, however, is still not very well understood, since different species, different environmental and geological settings and different socioeconomic conditions may affect the response of this service to climate change.

LAND TENURE AND OTHER FOREST RIGHT ISSUES

Deforestation and forest degradation in tropical and some of boreal forests are serious problems that contribute to the emission of greenhouse gases as well as to the fragmentation of forests. Deforestation and degradation have a series of direct and underlying causes, but none of these can be resolved if land and forest tenure are not clear or are not enforced. State land is more frequently subject to conversion into agricultural land than privately owned land.

Privately owned and concession forests, however, are increasingly coming under pressure, especially in countries with policies that recognize traditional rights or favour the rights of community inhabitants to their surrounding forests. In the Amazon region, community lands also receive increased pressure, possibly due to the regional infrastructural plans (IIRSA), speculation of future forest values under new international agreements on 8 climate change (REDD+), investment in bioenergy, the relatively large size of many community lands in relation to their population and the lack of financial and human resources to secure their borders.

Since many of the areas with land and forest rights concerns are in remote areas and refer to areas where people may have conflicting interests, regularizing these rights has been a major challenge in the past. Some progress has nevertheless been made in Latin America.

CHANGES IN POLICY ENVIRONMENT

REDD+ Expectations

Probably one of the more notable short-term changes in the policy arena is the discussion of GHG emissions reduction through REDD+ and management, conservation and restoration of forest carbon stocks. Large sums of money have been pledged against the demonstrable reduction of GHG emissions through REDD+, but so far, no international agreement has been reached on emissions reduction targets for developing countries. Further, in many pilot projects, measurable results have been interesting but financial benefits limited.

REDD+ expectations are manifold, depending on the interest group. Some of these expectations are justified, others not, and most are probably too ambitious. Implementation of REDD+ strategies will have to deal with most, if not all, of the challenges mentioned in this chapter.

At the same time it will require the implementation of a monitoring system, the extent and detail of which has not yet been agreed upon. While this has serious implications, the current (international and national) political environment is set to enable projects and countries alike, to meet at least some of these challenges. For the forest manager much of the challenge lies in adjusting management practices in favour of carbon accumulation, while at the same time maintaining biodiversity, recognizing the rights of indigenous people and contributing to local economic development.

Changes in Legislation

In Latin America, many countries implemented new forest legislation in the period between 1995 and 2000. While in some countries this was based on a thorough analysis of the forest sector, in others it was more in response to different pressure groups and based on changes in neighbouring countries.

In some countries (for example Costa Rica), new legislation was relatively successful in achieving the objective of forest conservation, although reducing forest use for timber production considerably (Louman, in print). In others, it has been difficult to implement new legislation if unaccompanied by other measures and if the process was not participatory and consultative.

More recently, countries have realized that they have better results when their new legislation is developed using more participative processes (for example in the DRC and Honduras). However, these processes are too young to be able to assess the true success in terms of increased implementation of legislative requirements.

Climate change will increase the challenge of designing and implementing new legislation that considers new international agreements, conflicts of interest in forest areas, as well as the need for coordination with other sectors. This may involve legislation on land and forest tenure, indigenous rights, the production of fuels and land use planning including restricting the access and use of certain areas or of some species, due to the risk of climate change impacts or the need of soil and water protection or maintenance of biological corridors.

In revising forest legislation, it is important to consider all related legislation, so that, for example, legislation or policies oriented at increasing forest area on private land is not nullified by policies or legislation that define forest land as 'un-used' or 'luxury possessions', taxing them relatively heavily or even threatening to expropriate the owners.

Social Responsibility Requirements

Concerns for sustainable development, for the deterioration of the environment and of social relations, as well as for the negative effects of climate change at different scales are influencing market decisions. This can above all be noticed in agricultural product markets, where buyers are looking for products that meet specific environmental and/or social standards. Some banana plantation owners that export to the European market, for example, have started to invest in forest land for conservation and carbon emissions compensation.

New standards have just recently been developed to monitor and evaluate carbon dioxide equivalent emissions from livestock farms in Costa Rica, while the COOPEDOTA coffee cooperative was recently declared carbon neutral.

These new developments pose interesting opportunities, more research is required to determine how these mechanisms can be used to improve the maintenance of other ecosystem services (such as water regulation and biodiversity maintenance), strengthen the adaptive capacity of natural and human systems and complement conservation and sustainable use of the existing forest areas within the agricultural landscapes.

Opportunity Costs of Land Use

Meeting REDD+ expectations has much to do with being able to identify the opportunity costs of local actors when they choose forest conservation and management rather than other land uses. Many of the REDD+ cost analyses are based on compensation for lost opportunities, although it has been found that forest conservation on private lands does not only occur for financial reasons.

If lands surrounding forests have high opportunity costs, there is the likelihood of increased pressure to convert those forests to the adjacent land use in order to make them more profitable. Opportunity costs may vary due to variations in market prices of the crops cultivated, government policies that subsidize agricultural inputs or the exportation of the outputs, or policies favouring the production of biofuel. The forest user or owner does not easily influence these factors.

As a group, in particular, if acting within the framework of REDD+, it may be possible to influence legislation, reduce the unequal treatment of forests as compared to agricultural crops, thus making forest management more competitive with other forms of land use.

UNCERTAINTY AND RISK MANAGEMENT IN FOREST CLIMATE

Climate change projections for the future involve a series of uncertainties. It is still not sure what emission scenario will best reflect reality, how these emissions change climate, in particular in relation to the distribution of precipitation or what other factors may play a role in influencing local vegetation and how local vegetation will react to climate and other factors. Thus, forest management for climate change has to deal with a range of uncertainties.

The challenge is to reduce those uncertainties and to design management systems that can deal with unexpected changes. Uncertainty and risk management options may involve monitoring systems (*e.g.* climate, biodiversity, production, and social impacts), early warning systems, working groups that analyze the implications of data obtained through monitoring, mechanisms dealing with risk of income loss, appeal systems for unpopular decisions as well as free prior and informed consent of indigenous and local communities. Flexible adaptive management approaches need to be a part of any management strategy that involves risk and uncertainty. Such strategies will need to include a set of tools, rather than one specific approach, to be able to switch from one to another tool, depending on local conditions, changes in those conditions, and success of already applied tools.

ECOLOGY MOVEMENTS AND CONFLICTS OVER NATURAL RESOURCES

The recent period in human history contrasts with all the earlier ones in its strikingly high rate of resource utilization. Ever expanding and intensifying industrial and agricultural production has generated increasing demands on the world's total stock and flow of resources. These demands are mostly generated from the industrially advanced countries of the North and the industrial enclaves in the underdeveloped countries of the South. Paradoxically, the increasing dependence of the industrialised societies on natural resources, through the rapid spread of energy and resource-intensive production technologies, has been accompanied by the spread of the myth that increased dependence on modern technologies implies a decreased dependence on nature and natural resources This myth is supported by the introduction of a long and indirect chain of resource utilisation which leaves invisible the real material resource demands of the industrial processes.

Through this combination of resource intensity at the material level and resource indifference at the conceptual and political levels, conflicts over natural resources generated by the new pattern of resource utilisation are generally shrouded and overlooked. These conflicts become visible when resource and energy-intensive industrial technologies are challenged by communities whose survival depends on the conservation of resources threatened by destruction and overexploitation, or when the devastatingly destructive potential of some industrial technologies is demonstrated as in the Bhopal disaster.

For centuries, vital natural resources like land, water and forests had been controlled and used collectively by village communities thus ensuring a sustainable use of these renewable resources. The first radical change in resource control and the emergence of major conflicts over natural resources induced by non-local factors was associated with colonial domination of this part of the world. Colonial domination systematically transformed the common vital resources into commodities for generating profits and growth of revenues. The first industrial revolution was to a large extent supported by this transformation of commons into commodities which permitted European industries access to the resources of South Asia.

With the collapse of the international colonial structure and the establishment of sovereign countries in the region, this international conflict over natural resources was expected to be reduced and replaced by resource policies guided by comprehensive national interests. However, resource use policies continued along the colonial pattern and, in the recent past, a second drastic change in resource use has been initiated to meet the international requirements and the demands of the elites in the Third World, leading to yet another acute conflict among the diverse interests. The most seriously threatened interest, in this conflict, appears to be that of the politically weak and socially disorganised group whose resource requirements are minimal and whose survival is primarily dependent directly on the products of nature outside the market system. Recent changes in resource utilisation have almost wholly by-passed the survival needs of these groups. These changes are primarily guided by the requirements of the countries of the North and of the elites of the South.

This book analyses environmental conflicts in contemporary human society. In general it relates to societies all over the world, but in particular it addresses the most intense and emerging social contradictions in India related to conflicts over natural resources. Science and technology are central to these conflicts because while scientific knowledge has been used by contemporary societies to considerably enlarge man's access to natural resources, it has also allowed the utilisation natural resources at extremely high rates.

The contemporary period is characterised by the emergence of ecology movements in all parts of the world which are attempting to redesign the pattern and extent of natural resource utilisation to ensure social equality and ecological sustainability. Ecology movements emerging from conflicts over natural resources and the people's right to survival are spreading in regions like the Indian subcontinent where most natural resources are already being utilised to fulfil the basic survival needs of a large majority of people.

The introduction of resource and energy-intensive production technologies under such conditions leads to economic growth for a small minority while, at the same time, undermines the material basis for the survival of the large majority.

In this way, ecology movements have questioned the validity of the dominant concepts and indicators of economic development. The ideology of economic development, which remained almost monolithic in the post World War II period, is thus faced with a major foundational challenge.

In this chapter an attempt has been made to provide a systematic conceptual framework for analysing the processes and structures of modern economic development from an ecological -perspective. It attempts to analyse the relationship between economic development and conflicts over natural resources to trace the roots of ecological movements.

Further, in the light of the ecological perspective, it examines the fundamental assumptions and categories of modern development economics that are used to determine the objectives of economic development as well as the criteria for the choice of technologies that are used to achieve these objectives.

ECONOMIC DEVELOPMENT AND ENVIRONMENTAL CONFLICTS IN INDIA

A characteristic of indian civilization has been its sensitivity to natural ecosystems. vital renewable natural resources like vegetation, soil and water were managed and utilised according to well defined social norms that respected the known ecological processes. The indigenous modes of natural resources utilisation were sensitive to the limits to which these resources could be used It is said that the codes of visiting important pilgrim centres Badrinath in the sensitive Himalayan ecosystem, included a maximum stay of one night so that the temple area would not put excess pressure on the local natural resources base. In the precolonial indigenous economic processes, the levels of utilisation of natural resources were not significant enough to result in drastic environmental problems. There were useful social norms for environmentally safe resource utilisation and people protested against the destructive use of resources even by kings. A major change in the utilisation of natural resources of India was introduced by the British who linked the resources of this country with the direct and large nonlocal demands of Western Europe. Natural resource utilisation by the East India Company, and later by the colonial rulers, replaced the indigenous organizations for the utilisation of natural resources, like water, forest and minerals, that were mainly managed as commons.

With the establishment of British colonial rule in India, the ever increasing resource demands of-the industrial revolution in England were largely met from colonies like India. Forced cultivation of indigo in Bengal and Bihar, cultivation of cotton in Gujarat and the Deccan led to large-scale commitment of land for the supply of raw materials for the British textile industry, the flagbearer of the industrial revolution. Forests in the sensitive mountain ecosystems like the Western Ghats or the Himalayas were felled to build battleships, or to meet the requirements of the expanding railway network. Forests of the Bengal-Bihar-Orissa region were used for running wood fuel locomotives in the early stages of railway expansion.

The latter stages of colonial resource utilisation and control included the monopolization. of water rights as in the Sambhar Lake of Rajasthan or the Damoda' Canal in Bengal. Colonial intervention in natural resource management in India led to conflicts over vital renewable natural resources like water or forests and induced new forms of poverty and deprivation. Changes in resource endowments and entitlements introduced by the British came into conflict with the local people's age old rights and practices related to natural resource utilisation As a result local responses were generated through which people tried to regain and retain control over local natural resources. The indigo Movement in Eastern India, the Deccan Movement for land rights or the forest movement in all forest areas of the country, the Western Ghats, the Central Indian Hills or the Himalayas, were obvious expressions of protest generated by these newly created conflict's. Conflicts generated by the colonial modes of natural resource exploitation could not, however, grow with a local identity. With the progress of the anti-colonial people's movement at the national level, these local protests merged with the national struggle for independence. With the collapse of colonial rule internationally, and the emergence of sovereign independent countries in the Third World like India, resolution of these conflicts at the local level became a possibility. While political independence vested the control over natural resources with the Indian state, the colonial institutional framework for natural resource management did not change in essence. Where colonialism collapsed, the slogan of economic development stepped in. There was unfortunately no alternative institutional mechanism other than that of the classical model of development left by the British, with which the newly formed Indian state could respond to the accentuated aspirations of the Indian people for a better life.

The same institutions and concepts, nurtured and developed by the colonial rulers were applied to objectives which were exactly opposite to those of the colonial period. Concepts and categories relating to economic development and natural resource utilisation that had emerged in the specific context of capitalist growth and industrialization in the centres of colonial power were raised to the level of universal assumptions and applicability. The processes which led to deprivation were now entrusted with the responsibility of basic needs satisfaction. No serious thought was given to the fact that the historical specificity of early industrial development in Western Europe necessitated the permanent occupation of the colonies and the undermining of the local 'natural economy'.' This inexorable logic of resource exploitation, exhaustion and alienation integral to the classical model of economic development based on resource intensive technologies led Gandhi to seek an alternate path of development for India when he wrote:

God forbid that India should ever take to industrialism after the manner of the West. The economic imperialism of a single tiny island kingdom (England) is today keeping the world in chains. If an entire nation of 300 million took to similar economic exploitation, it would strip the world bare like locusts.

While Gandhi's critique was a forewarning against the problems likely to arise by following the classical path of resource-intensive development, at the time of India's independence, there was no clear and comprehensive work plan to realise the Gandhian dream of alternate development that would be resource prudent and would satisfy basic needs. The issues of resource constraints of economic development were, therefore, not highlighted at the theoretical level, partly due to the tremendous pressure of the enhanced developmental aspirations of a newly independent nation, and partly due to the lack of internalization of natural resource parameters within the framework of economics. As the scale of economic development activities escalated from one

Five Year Plan to another, the disruption of ecological processes that maintain the productivity of the natural resource base started becoming increasingly apparent.

The classical model of economic development in the case of the newly independent nations resulted in the growth of urban-industrial enclaves where commodity production was concentrated, as well as rapid exhaustion of the internal colonies whose resources supported the enhanced demands of these enclaves. In the absence of ecologically enlightened resource management methods, the pressure of poverty enhanced the pace of economic development activities in the hope of a quick improvement in the standard of living for all, as in the case of Western Europe. For example, commercial forestry earned more revenue by making increasing amount of timber and pulpwood available in the market but in the process reduced the multipurpose biomass productivity or damaged the hydrology of the forests. People dependent on non-timber biomass outputs of forests like leaves, twigs, fruits, nuts, medicines and oils were unable to sustain themselves, in the face of the commercial exploitation of forests. The changed hydrological character of the forests affected both the micro-climate and the stream flows, disturbing the hydrological stability and affecting agricultural production.

There are similar examples from all parts of the country, related to almost all massive developmental interventions in India's natural resource system. Ecological degradation and economic deprivation generated by the resource insensitivity and intensity of the classical model of development have resulted in environmental conflicts, an understanding of which is imperative for the reorientation of our current development priorities and concepts. It is becoming in creasingly clear that these classical concepts and priorities are being used as an alibi to direct 'development' at the national level, while the educated minority elite is the main beneficiary of these 'development' processes.

The ecology movements that have emerged as major social movements in many parts of India are making visible many invisible externalities and pressing for their internalisation in the economic evaluation of the elite-oriented development process. In the context of a limited resource base and unlimited development aspirations, ecology movements have initiated a new political struggle for safeguarding the interests and survival of the poor, the marginalised, including women, tribals and poor peasants.

Ecology Movements and Survival

The intensity and range of ecology movements in independent india have continuously widened as predatory exploitation of natural resources to feed the process of development has increased in extent and intensity. This process has been characterised by the massive expansion of energy and resource-intensive industrial activity and major development projects like large dams, forest exploitation, mining and energy-intensive agriculture. The resource demand of development has led to the narrowing of the natural resource base for the survival of the economically poor and powerless, either by direct transfer of resources away from basic needs or by destruction of the essential ecological process that ensure renewability of the life-supportirig natural resources.

In the light of this background, ecology movements emerged as the people's response to this new threat to their survival and as a demand for the ecological conservation of vital life-support systems. The most significant life-support systems in addition to clean air are the common property resources of water, forests and land on which the majority of the poor people of India depend for survival. It is the threat to these resources that has been the focus of ecology movements in the last few decades. Among the various ecology movements in India, the Chipko movement (embrace the trees to oppose fellings) is the most well known. It began as a movement of the hill people in the state of Uttar Pradesh to save the forest resources from exploitation by contractors from outside.' It later evolved into an ecological movement that was aimed at the maintenance of the ecological stability of the major upland watersheds in India. Spontaneous people's response to save vital forest resources was seen in Jharkhand area in Bihar-Orissa border region as well as in Bastar area of Madbya Pradesh where there were attempts to convert the mixed natural forests into plantations of commercial tree species, to the complete detriment of the tribal people. In the southern part of India the Appiko movement, which was inspired by the success of the Chipko movement in the Himalayas, is actively involved in stopping illegal over-felling of forests and in replanting forest lands with multipurpose broad leaved tree species. In Himachal Pradesh the Chipko activists have intensified their opposition to the expansion of monoculture plantation of the commercial Chir Pine (Pinus roxburghii). In the Aravalli Hills of Rajasthan there has been a massive programme of tree planting to give employment to those hands which were hitherto engaged in felling of trees.

The exploitation of mineral resources, in particular the opencast mining in the sensitive watersheds of the Himalayas, the Western Ghats and Central India have also resulted in a great deal of environmental damage. As a consequence, environmental movements have come up in these regions to oppose the reckless mining operations. Most successful among them is the movement against limestone quarrying in the Doon Valley. Here, volunteers of the Chipko movement have led thousands of villagers, in peaceful resistance, to oppose the reckless functioning of limestone quarries that is seen by the people as a direct threat to their economic and physical survival.'

While the Doon Valley instance has a long history of popular opposition to the quarrying of limestone and a Supreme Court order has restricted the area of quarrying to a minimum, examples of such success' of ecology movements are rare People's ecology movements against mineral exploitation in the neighbouring areas of Almora and Pithoragarh still seem to be ignored, probably due to the relative isolation of these interior areas. Beyond the Himalayas, the ecology movement in the Gandhamardan Hills in Orissa against the ecological havoc of bauxite mining has gained momentum and it draws inspiration from the Chipko movement.

The mining project of the Bharat Aluminium Company (BALCO) in the Gandhamardan Hills is being opposed by local youth organisations and tribal people whose survival is directly under threat. The peaceful demonstrators have claimed that the project could be only continued 'over our dead bodies. The situation is more or less the same in large parts of Orissa-Madhya Pradesh region where rich mineral and coal deposits are being opened up for exploitation and thousands of people in these interior areas are being pushed to deprivation and destitution. This is also true of the coal mining areas around the energy capital of the country in Singrauli. In these interior areas of Central India, movements against both mining and forestry are becoming increasingly volatile and people's resistance is growing.

Large river valley projects, which are coming up in India at a very rapid pace, is another group of development projects against which people have organised ecology movements. The large-scale submersion of forest and agricultural lands, a prerequisite for the large river valley projects, always takes a heavy toll of dense forests and the best food growing lands. These have usually been the material basis for the survival of a large number of people in India, specially tribal people.

The Silent Valley project in Kerala was opposed by the ecology movement on the ground of its being a threat, not to the survival of the people directly, but to the gene pool of the Tropical Rainforests threatened by submersion. The ecological movement against the Tehri high dam in the UP Himalaya exposes the possible threat to people living both above and below the dam site through large-scale destabilization of land by seepage and strong seismic movements that could be induced by impoundment.

The Tehri Dam Opposition Committee has appealed to the Supreme Court against the proposed dam by identifying it as a threat to the survival of all people living near the river Ganga up to West Bengal. Most notable among the people's movements against dams on the issue of direct threat to survival from submersion are Bedthi lcchampalli, Bhopalpatnam, Narmada Sagar, Koel-Karo, Bodhghat, etc.

In the context of the already overutilised land resources, the proper rehabilitation on a land-to-land basis of millions of people displaced through the construction of dams seems impossible. The cash compensation given instead is inadequate in all respects for providing an alternate livelihood for the majority of the displaced. Destitution is thus the first and foremost precondition for initiating large dam projects

While the process of construction of dams itself invites opposition from ecology movements, the functioning of water projects dependent on the constructed dams results in further ecological disasters and movements. People's movements against widespread water-logging, salinisation and the resulting desertification in the command areas of many dams have been registered. Among them are instances of protests against the Tawa, Kosi, Gandak, Tungabhadra, Malaprabha, Ghatprabha projects and the canal irrigated areas of Punjab and Haryana. While excess water led to ecological destruction in these cases, improper and unsustainable use of water in the arid and semi-arid regions generated ecology movements in a different way. The anti-drought and desertification movement is gaining momentum in the dry areas of Maharashtra, Karnataka, Rajasthan, Orissa, etc. Ecological water use for survival is being advocated by water based movements like Pani Chetana, Pani Panchayat, and Mukti Sangharsh. Another major movement originating from the ecological destruction of resources by growth based development is spreading all along the 7,000 km long coastline of India. It is the movement of the small fishing communities against the ecological destruction caused by mechanised fishing whose instant profit motive is destroying the coastal ecology and its long-term biological productivity in a big way.

No amount of threat to survival in India from environmental hazards can be complete without a reference to the Bhopal tragedy on 2 December 1984, in which several thousand people died and several lakhs faced serious health hazards following the leakage of poisonous Methyl Iso Cyanate from a pesticide plant of Union Carbide (India) Limited. People's movements for clean air and water are growing in ail parts of the country just as ecologically irresponsible industrialization is moving deeper into the hinterland in search of new resources.

ACHIEVE SUSTAINABLE FOREST MANAGEMENT IN INDIA

Historically, Indian forest policies have alienated people from the forests, thereby, exacerbating the rates of deforestation. Post-independence forest policies contributed to an expansion in agricultural production, met industrial demand for raw materials, and tightened control of forest lands through restricted access to forests and forest products. Protection policies increased the hardships of vulnerable social groups by denying them access to forests.

While the state took responsibility for managing forest resources, it did not have the commensurate resources to effectively manage and police the forests from traditional users. Before state intervention, forests were managed as communal property; the crucial role of forests in the economic subsistence of individuals, families and community was the basis for managing them as communal resources. A failure to recognize community control of forests led to a collapse in institutional norms that were instrumental in protecting and managing forest resources for local use. A shift in property rights to the state steadily undermined the rights of tribals to use and extract forest resources. Involvement of rural communities living close to forests in protection and management of forest resources is enshrined in the National Forest Policy 1988.Translation of policy found expression in the resolution of Government of India, Ministry of Environment and Forests issued in June 1990.It envisaged that in lieu of the participation, the local communities will be entitled to sharing of usufructs in a manner specified by the concerned State Forest Departments. This led to the initiation of Joint Forest Management (JFM) programme. Importance of the programme is evident from the fact that the Government of India has constituted a "JFM Network" with the Inspector General of Forests, Government of India as the Chairman. The objectives of the network are (i) to act as a regular mechanism of consultation between various agencies engaged in JFM work in the country and (ii)To obtain constant feedback from various stakeholders on the JFM programme for proper policy formulation and suitable direction to States.

World leaders adopted the Millennium Declaration at the Millennium Summit in September 2000. The proportion of land area covered by forest globally is one of the indicators for the seventh MDG i.e to ensure environmental sustainability.In addition to quantitative, time-bound targets, the Millennium Declaration calls for other actions, including intensified efforts for "the management, conservation and sustainable development of all types of forests", an international commitment to sustainable forest management made in 1992 at the United Nations Conference on Environment and Development (UNCED) and embodied in the Forest Principles. Subsequent intergovernmental deliberations to promote progress towards sustainable forest management took place in the Intergovernmental Panel on Forests (IPF) and Intergovernmental Forum on Forests (IFF) from 1995 to 2000, and continue in the United Nations Forum on Forests (UNFF) as well as in other fora.

FORESTS AND THE MILLENNIUM DEVELOPMENT GOALS

The MDGs call for the integration of the principles of sustainable development into environmental policies. Environmental sustainability is being mainstreamed in forest policies around the world, particularly since UNCED, while the integration of the goals of poverty and hunger reduction in forest policies and plans is less widespread.

Community-based forestry, or participatory forestry, is particularly well placed to address poverty reduction. Community-based forestry is now well accepted and established in various countries in all regions, and programmes are beginning to generate financial and other benefits. Improving local peoples' rights and access to forest resources is a tool to the success of community-based forestry programmes. However, much still remains to be done to clarify and secure access rights. Many countries are working to strengthen forest governance, some through decentralization processes that allow the poor to derive more benefits from forests and be more involved in decision-making and forest management itself.

Although improving rights and access to forest resources and developing small-holder forest-based enterprises (including through community-private sector partnerships) show particular promise for poverty reduction, local political and economic realities, opportunity costs for the use of local resources, and other factors may prevent the poor from benefiting from community-based forestry programmes to the extent intended.

Intersectoral coordination is important for the achievement of all MDGs, but is particularly critical for reducing poverty and hunger and ensuring environmental sustainability, which are highly cross-sectoral by nature. Improved intersectoral cooperation and coordination will help efforts both to integrate the principles of sustainable development into forest-related policies and to integrate forests into sustainable development plans.

Joint Forest Management

Forest-based poverty reduction efforts tend to be linked to other land uses and should form a part of rural development strategies. Conversely, the potential for forests and trees outside forests to contribute to environmental sustainability cannot be fully realized without intersectoral cooperation and coordination. Intersectoral coordination, although difficult and time consuming, is necessary for sound decisions on land use and resource allocation, particularly when there are trade-offs between national development goals.

National Forest Policy in India treats forests as environmental and social resource. With the initiative of assigning ownership of Non Timber forest Produce (NTFP) to the local communities including the grass root level democratic institution for enhancing their livelihood opportunities and also improving their income with the value addition.

India has shifted the approach of forest management from regulatory to participatory mode of management with the resolution promulgated in 1990. At present, more than 17 million forests is managed by almost 10,000 Joint Forest Management Committees with the benefit sharing mechanism. In addition, the Government of India is in process to frame legislation for the settlement of tenurial rights of the forest dwelling communities mainly tribal on forests. This would definitely help in reducing the poverty of forest dwelling communities. The JFM resolution was circulated by Ministry of Environment and Forest in the year 1990 and 2000. JFM is a government resolution. A government resolution is a executive order or opinion of the legislature. A resolution does not have any legal backing.JFM as the term indicates is the management of forest by more than one party. In India there are two parties: the government represented by the Forest Department and the people living in villages located within forest or on the fringes. There are two major reasons behind introducing JFM: one that the government's management system was not succeeding in arresting growth of forest degradation and deforestation.Second a new management paradigm was evolving in which the local people's participation was found to be an appropriate and promising tool in arresting forest degradation.

However in pre-independent India the concept of JFM didn't exist. The first National Forest Policy was adopted in 1894. Following were the guide lines

- Ensure maintaneance of adequate forest cover
- Meet the needs of local people.
- Collect maximum revenue after meeting the needs of the local people.
- Give priority to permanent cultivation over forestry land

In post independent India there was a shift in policy. In1988, the new forest policy was adopted which covered all the sustainable management approaches. The new policy had a few unique features. Which were as follows;

- Maintenance of environmental stability and restoration of ecological balance, soil and water conservation
- Conservation of natural heritage and genetic resources.
- Increasing productivity to meet the local needs then the national need
- Creating massive peoples participation movement to protect forest and tree cover and achieve the objective of reducing pressure on existing forests and meeting peoples need.
- Deriving economic benefits must be subordinated to these principal aims.

This initiated a process of reform at the local policy and operational level of forest management ensuring that the Forest Department developed close collaboration for protection and sustainable management of forests.

The aim was –Involvement of village communities and voluntary agencies of degraded forest land.

Important guidelines were as follows:

- The programme should be implemented under an arrangement between a voluntary agency or beneficiaries and the State Department.
- No ownership rights or lease should be given over the forest land.
- The beneficiaries should be entitled to share usufructs to the extent and subject to conditions prescribed by the State Government.
- Access to forest land usufructs should be available only to benefactress who get organized into a village institution especially for forest regeneration and protection. This could be through a village panchayat or a Village Forest Committee.
- The beneficiaries should be given usufructs like grass, lops and tops

of branches and minor forest produce. If they successfully protect the forest they will be a portion from the sale proceeds when they mature.

- Areas selected from the programme should be free from claims from any person who is both a beneficiary under the scheme.
- The selected site should be worked in accordance of Working Scheme duly approved by the state government.Such a scheme may remain in operation for ten years and revised after that. The working scheme is prepared in consultation of with the beneficiaries
- It should ensure that there is no grazing at al on the forest land protected by the Village Forest Committee. Permission to cut and carry grass free of cost should be given so that stall feeding is promoted.
- No agriculture should be promoted on the forest land.
- Cutting of tress should not be permitted before they are ripe for harvesting. The Forest Department should not be permitted to cut to cut trees protected by the Village Forest Committee except in a manner prescribed in the working scheme. In case of emergency needs the village community should be taken into confidence.
- The Forest Department should closely supervise the work.If beneficiaries are unable to perform their assigned duties in a satisfactory manner the usufctory benefits will be withdrawn without giving any compensation.

Such set-up, however suffered from certain flaws:

- Bye- laws have not been formulated for the functioning of the JFMC though now most of the states have issued executive orders for the functioning of JFMC but these executive orders are not binding on JFMC. The aim was to decentralize the process and make JFMC.
- Minor Forest Produce (MFP) has not been defined neither by the State legislature or by the Centre.
- Central and State/UT Governments have issued guidelines for the creation and functioning of JFMC but these guidelines are not in conformity with the provisions of the Constitution
- Through the Constitution (Seventy-third Amendment) Act, 1992 and Panchayat (Extension to the Scheduled Areas) Act, 1996 ownership rights over minor forest produce (MFP) have been given to Village Panchayats. Now on the same resource base *i.e.* NTFP/MFP we have two sets of groups having ownership and while the Panchayats have a legal backing, JFMC don't have. conservation have been raised while the proponents of Panchayats have termed JFM as a parallel institution against the spirit of the Constitution.
- The JFM resolution only provides 20% of share to be given the Joint

Forest Management committees, while rest of the income would go to the Forest Department. There is an obvious unequal distribution of benefit sharing between the parties. These committees would have to protect the forest for ten years,and would only receive 20% of the share.

- The national resolution provides that at least 33% of the seats shall be reserved for women in the specified committees. However states like Rajasthan are not following this provision.

Panchayati Raj System

The concept regarding Panchayati is contained in the Part IX of the 73rd amendment of The Indian Constitution.This amendment came into force in on 24th April 1993. It institutionalized the third stratum of government (Panchayats) at the local level.

Through the Constitution (Seventy-third Amendment) Act, 1992) Part IX "The Panchayats" was inserted in the Constitution which paved the way for "Village Panchayats" by making provisions for the constitution of Panchayats, their composition, election, powers, authority, responsibility, audit, etc. This Act doesn't apply to Schedule Area referred to in clause (1) and tribal areas referred to in clause (2) of Article 244 and certain other specified areas Village Panchayats have been given the responsibility of social/farm forestry, minor forest produce (MFP), and soil conservation through Eleventh Schedule.

"Panchayat" means an institution (by whatever name called) of self-government constituted under article 243B, for the rural areas. The amendment it fails to define what is 'self government'. It does not clarify whether self government means complete autonomously or extension of the state.It is left to states to derive their own interpretation regarding the nature of self government at the local level.By the word "nature" of self government it is meant reservation to be provided for the marginalized section of the society, the powers given to each level of self governing relating to administration, execution and performing the defined roles would be different in each State. Thus this will defeat the purpose of uniformity regarding the nature and concept of self government at an all India level.

The flaw again lies that the amendment has failed to assert the nature of 'Power And Authority'.This task is left to the state legislatures.Again defeating the purpose of uniformity. The consequences will be such that either the states will deny power to the panchyats instead of becoming autonomous local governing units they will basically remain in bureaucratic control.The burecratic trend can be seen most of the states.It is seen in the State of Goa power and functions have not yet been transferred to the Panchayats. As per the law there should be continuous elections every five years.However the statistics says that this status is not achieved in most of the states.None of the states except

WestBengal, Tripura and Rajasthan have not been able to hold continuous elections every five years. The judgment provided by the Supreme Court states that it is mandatory for every state to hold elections to Panchayat.

Giving powers to the third stratum of government for effective management of resources at the district,intermediate and village level. Parallel institutions have come about for management of the natural resources which

Participatory Approach

According to United Nations Development Programme's report 67.7 million people belonging to "Scheduled Tribes" in India are generally considered to be 'Adivasis', literally meaning 'indigenous people' or 'original inhabitants', though the term 'Scheduled Tribes' is an administrative term used for purposes of 'administering' certain specific constitutional privileges, protection and benefits for specific sections of peoples considered historically disadvantaged and 'backward'.

Out of the 5653 distinct communities in India, 635 are considered to be 'tribes' or 'Adivasis'. With the ST population making up 8.08% (as of 1991) of the total population of India, it is the nation with the highest concentration of 'indigenous peoples' in the world. 68 million tribals who inhabit forests and wild lands throughout India, Adivasis have evolved an intricate convivial-custodial mode of living.

The Beginning of Land Alienation

Introduction of the alien concept of private property began with the Permanent Settlement of the British in 1793 and the establishment of the "Zamindari" system that conferred control over vast territories, including Adivasi territories, to designated feudal lords for the purpose of revenue collection by the British. This drastically commenced the forced restructuring of the relationship of Adivasis to their territories as well as the powe relationship between Adivasis and 'others'. The predominant external caste-based religion sanctioned and practiced a rigid and highly discriminatory hierarchical ordering with a strong cultural mooring.

After the transfer of power, the rulers of the Residency Areas signed the "Deed of Accession" on behalf of the ruled on exchange they were offered privy purse. No deed was however signed with most of the independent Adivasi states. They were assumed to have joined the Union. The government rode rough shod on independent Adivasi nations and they were merged with the Indian Union. This happened even by means of state violence as in the case of Adivasi uprising in the Nizam's State of Hyderabad and Nagalim. The Constitution of India, which came into existence on 26 January 1950, prohibits discrimination on grounds of religion, race, caste, sex or place of birth and it provides the right to equality, to freedom of religion and to culture and education. STs are supposedly addressed by as many as 209 Articles and 2 special schedules of the Constitution - Articles and special schedules which are protective and paternalistic. Article 341 and 342 provides for classification of Scheduled Castes (the untouchable lower castes) and STs, while Articles 330, 332 and 334 provides for reservation of seats in Parliament and Assemblies. For purposes of specific focus on the development of STs, the government has adopted a package of programmes, which is administered in specific geographical areas with considerable ST population, and it covers 69% of the tribal population.

Forest Laws in Relation to Tribal Land Conflict

The total forest cover in India is reported to be 765.21 thousand sq. km of which 71% are Adivasi areas. Out of this 416.52 and 223.30 thousand sq. kms. is categorised as reserved and protected forest respectively. About 23% of these are further declared as Wild Life Sanctuaries and National Parks which alone has displaced some half a million Adivasis. By the process of colonisation of the forests that began formally with the Forest Act of 1864 and finally the Indian Forest Act of 1927, the rights of Adivasis were reduced to mere privileges conferred by the state.

The Imperial Forest Department was formed in 1864. The first Act for the regulation of forests was passed in 1865. It empowered the government to declare any land covered with trees or brushwood as government forest and to make rules to manage them. The act was applicable only to the forests in control of the government and did not cover private forests. It made no provision regarding the rights of the users. Its primary function was to generate revenue by selling timber and forest produce. Since forests are deemed to be government property, Adivasis were barred from collecting forest produce.

According to the Indian Forest Act, 1927, the Government can constitute any forest land or waste land which is the property of Government or over which the Government has proprietary rights, a reserved forest, by issuing a notification to this effect. Commercial interests of the then British Government motivated it to declare more and more lands as reserved forests, without ascertaining the rights of the tribals and other forest dwellers.

After the enforcement of the Indian Forest Act in 1927, the government declared all lands not claimed by private individuals and agencies as forest lands and classified them into reserve, protected and village forests. To attract labour to work in forestry related activities, the governments gave lands to households for cultivation and also promised jobs. A number of forest villages were established. The lands allotted for cultivation were on a purely temporary basis and forest officials were to look after the administration of these villages After Independence, the inhabitants of forest villages realized that most villages lacked infrastructure facilities; they could not get loans for development of agriculture as they had no title over the land. Around 1980, the government decided to convert forest villages into revenue villages. This was done in many states. However, not all the forest villages have been converted into revenue villages. In some villages, land has been given on a 15 year lease to enable the holders to obtain loans from banks. Residents have agitated for the conversion of these villages into revenue villages. Even after independence in 1947, during the process of amalgamation of princely states, the activity of consolidation of government forests continued. The State Governments/ UT Administrations proclaimed the lands of ex-princely states and the zamindar-lands as Reserved Forests. However, no effective steps were taken to give an alternative accommodation to the tribal people residing in this area.

The Act invites claims for settlement in the reserve forest area. This is provided from section 3 to section 20 of the IFA. These sections provide a complicated procedure which requires documentation in relation to ownership of the land, evidence, which would be impossible for an indigenous to produce in front of the Settlement Officer. Many tribal families today are regarded as "encroachers" on forest land (persons who are cultivating or residing on forest land without a legal right to be there).

In many cases, however, it would be more accurate to say that the "forest" is encroaching on their traditional rights. There are a number of core issues that are central to the resolution of forest based conflicts, and are integral to the question of encroachment.

Protected Areas

About 4.5% of the total land in the country is covered under protected areas, classified into national parks, wildlife sanctuaries and protected areas. The Wildlife (Protection) Act of 1972 provided for the administration of these areas. The act places several restrictions on the residents of villages in these areas. Efforts are also being made to relocate these villages outside such areas. These efforts have met with stiff resistance from the residents and violent conflicts have taken place in many places. Under pressure from the World Bank and the environment protection fund agencies, the government has tightened the restrictions on the hunting of animals included in the list of wildlife species, as also a number of economic activities. Consequently, development projects in these areas have been shelved.

Throughout this, one needs to be acutely aware that there are huge forest encroachments by powerful vested interests which need to be dealt with strictly. Mechanisms will need to be put in place to ensure that powerful interests/ land mafias do not use poor people as a front to further their encroachments. In the past two decades even as the genuine forest dwellers have lived in the fear of eviction, mafias have continued to encroach forests and alienate tribal land. This has not only had severe ecological implications due to disturbance of forest and wildlife but also displaced tribals from their homes and livelihoods.

Forests Conservation Act

The 42nd Constitutional Amendment shifts forests from the "State List" to the "Concurrent List". The FCA prohibits non-forest use of forest land without the government's approval. It also advocates "sustainable forest management through participatory approach", with "due regard to the traditional rights of the tribal people on forest land". Further making the claim settlement process stringent and inflexible and thereby the settlement dispute continues for decade. The 1990 guidelines on Tribal and Forest Interface expressly provides for speedy settlement of claims in relation to forest land, Regularization of Encroachments; Review of Disputed Claims over Forest Land; Regularization of Pattas & Leases; and Conversion of Forest Villages to Revenue Villages. These Guidelines were supposed to provide a framework to resolve the problem of settlement of rights of tribals and other forest dwellers on forest land. For instance, FP1 outlined procedures whereby state governments could apply to the Government of India for regularization of "pre-1980 encroachments". However, these Guidelines have not been implemented

On 5 February 2004, the MoEF issued new guidelines for "regularization of the rights of the tribals on the forest lands", in continuation of the 1990 Guidelines. These "supplementary guidelines" request the State Governments to give legal recognition to "the traditional rights of the tribal population on forest lands", and to submit proposals for conferring "heritable but inalienable rights over such lands" on "tribal dwellers who are in continuous occupation of such forest land at least since 31/12/1993".

There are thousands of cases of local inhabitants claiming that they were in occupation of notified forestlands prior to initiation of forest settlements under the Indian Forest Act. There are a number of cases of pattas/leases/grants said to be issued under proper authority but which has now become contentious issues between different departments, particularly the Forest Department and the Revenue Department. The problem is compounded by the fact that in many cases there is no clear demarcation of forest lands. In fact most of the disputes and claims relating to use and access to forests have lingered on and evaded resolution in the past because of the failure to demarcate precisely the extent of the forest. All of these require remedies and an approach aimed at only evicting the forest-dwellers is worsening the situation, not remedying it.

There is a need to consciously distinguish between the authorised/ unauthorized occupation of forests by dwellers for bonafide consumption, and the use of forest resources from the large scale timber and fuel wood smuggling and intensive commercial exploitation of forest resources. In the former case the so-called encroachment is a local and subsistence-oriented activity, while in the latter case it is a widespread organized industry largely driven by the mafia. The approach of the law and policy to the two situations cannot be the same

The National Forest Policy was adopted in 1894. Following were the guide lines:

- Ensure maintenance of adequate forest cover
- Meet the needs of local people.
- Collect maximum revenue after meeting the needs of the local people.
- Give priority to permanent cultivation over forestry land.

In post independent India there was a shift in policy. The National Forest Policy Resolution1952 adopted by the government, emphasised that the forest policy should be based on paramount national needs. For the first time the resolution emphasized the ecological and social aspects of forest management, giving secondary importance to the needs of commerce, industry and revenue. The resolution did not call for any change in the forest law and remained only a pious declaration.

The National Forest Policy, 1988, has a separate section on 'Tribal people and forests' which states that: "having regard to the symbiotic relationship between the tribal people and forests, a primary task of all agencies responsible for forest management, including the forest development corporations should be to associate the tribal people closely in the protection, regeneration and development of forests as well as to provide gainful employment to people living in and around the forest." The section on 'rights and concessions' has focussed on 'full protection' of the rights of tribals.

In1988, the new forest policy was adopted which covered all the sustainable management approaches. The new policy had a few unique features. Which were as follows;

- Maintenance of environmental stability and restoration of ecological balance, soil and water conservation
- Conservation of natural heritage and genetic resources.
- Increasing productivity to meet the local needs then the national need
- Creating massive peoples participation movement to protect forest and tree cover and achieve the objective of reducing pressure on existing forests and meeting peoples need.
- Deriving economic benefits must be subordinated to these principal aims.

This initiated a process of reform at the local policy and operational level of forest management ensuring that the Forest Department developed close collaboration for protection and sustainable management of forests.. While the resolution adopted a pro-tribal policy, the old Act of 1927 with all the subsequent amendments remained unchanged. The Scheduled Tribes and Other Traditional Forest Dwellers (Recognition of Forest Rights) Act, 2007.

The recently notified Act aims to compensate the "historical injustice" done to forest-dwelling tribes that were denied their traditional rights to forest lands and resources in the last couple of hundred years. In this period, the lands they had been dwelling on were declared "forest land" or "protected areas" for wildlife. The traditional conservation ethos and dependence of tribal societies on forest land were overlooked. The Act aims to reverse the alienation of tribes from their own habitat caused by past policies and laws.

The rights proposed to be accorded to Adivasis include pattas to forest lands occupied before 1980. This is not new; it has been a stated policy of the government for years now. A number of other rights are also to be granted: nistar (usufruct) or ownership rights to forest resources, grazing rights including seasonal ones of nomadic communities, habitation rights (for those classified as Primitive Tribal Groups), conversion of forest villages into revenue villages, and so on. No tribal person is to be evicted from currently occupied land until the process of determining rights is completed.

Under the Act:

- All rights are accompanied by responsibilities for forest protection;
- All right holders have the duty to conserve forests and wildlife, protect catchment areas, water sources, and ecologically sensitive areas, and inform the gram sabha of ecologically destructive activities;
- Hunting is explicitly excluded from the list of forest rights;
- All rights are meant only for bona fide livelihood needs, not for exclusive commercial purposes;
- In no case would forest land beyond 2.5 hectares be allotted (only land under occupation prior to 1980 can be considered, no new forest lands would be allotted).

The MOEF is of the view, land settlement of the forest dwellers would increase the problem of encroachment to the already depleting forest cover in our country, poaching would be rampant.

Case Laws

One of the earliest cases relating to forest rights was that of Banwasi Seva Asharam v. State of Uttar Pradesh. The case was initiated as a public interest writ petition under article 32 of the Constitution on behalf of the local people protesting for reservation of forest land by the state. People in the village have lived for generations and had relied for their sustenance on the forest products. The petitioners had alleged that state has ignored their claims of these people over the forest and steps were being taken foe eviction of many forest dwellers. A high level committee was formed by the apex court to look into the matter of claims.

In the meantime, the government informed the court it wanted to site National Thermal Power Station over the disputed forest land. The land was granted by the by the court to NTPC, thereby ordering eviction of the forest dewellers. The reasoning given by the court that forests are as much national assets, however generation of electricity would be equally important.The court did not address the primary aim of claim of the forest dwellers arising out of their dependence on the forest life guaranteed by article 21 of the Indian Constitution

The court dropped the all the claims over the land which the state has notified as a reserved forest area. The claims of the petitioner were written off on procedural grounds. The written application were not stamped, oral objections would not be considered, the tribal were ignorant of their legal rights

The NTPC asked to rehabilitee the tribal by housing transportation employment, infrastructure. In T.N Godavarman Thirumulkpad v. Union of India, the apex court issued directives sweeping directives to oversee the enforcement of forest laws across the nation. The court froze all forest wood based industries, enforced the scope of embargo on forest exploitation, and issued directions for sustainable forest management. The court focused the issues affection North East region where wood based industry contributes sustainability of the regions economy.

The issue raised by the amicus curiae was whose interest should be protected, the forest dweller or forest persons whose livelihood depends upon?The court stated that in depth examination of the Forest policy should be taken care off.

It should conform to the directions of National Forest Policy 1988. As stated in the policy the respective department of the states should encourage tribal welfare and joint forest management.

International Perspective

The right of indigenous peoples have been recognized by international agencies and organizations. Every individual has right to have a living, proper housing, drinking water, education, electricity, health facilities, communication, right over the land, they are denied everything. United Declaration of Human Rights(1948) and International Covenant on Civil and Political Rights (1966) have specific significance for indigenous peoples. The Universal Declaration provides a common standard for the human rights of all peoples and all nations, and proclaims the importance of traditional, political, and civil rights, as well as basic economic social and cultural rights e]veryone has the right to own property alone as well as in association with others." Furthermore, "no one shall be arbitrarily deprived of his property. The Covenant spells out civil and political rights and guiding principles based on the Universal Declaration.

The 1957 International Labour Organisation Convention, Protection and Integration of Indigenous and Other Tribal and Semi-Tribal Populations in Independent Countries, addresses the right of indigenous peoples to pursue material well-being and spiritual development, and was a first international instrument in specific support of indigenous peoples. Largely because of its view that indigenous peoples should be integrated into the larger society, a view that subsequently came to be seen by many as inappropriate. It was followed by Convention Concerning Indigenous and Tribal Peoples in Independent Countries. It presents the fundamental concept that the way of life of indigenous and tribal peoples should and will survive, as well as the view that indigenous and tribal peoples and their traditional organizations should be closely involved in the planning and implementation of development projects that affect them. The convention specifically provides recognition of traditional land rights of the indigenous people. The concentration gives the indigenous people the right to sustain themselves through natural resources, the collection of forest produce and speedy recovery of traditional rights of the indigenous people.

India has signed convention on Protection and Integration of Indigenous and Other Tribal and Semi-Tribal Populations in Independent Countries, but not Convention Concerning Indigenous and Tribal Peoples in Independent Countries.

The 1992 Convention on Biodiversity calls on contracting parties to respect traditional indigenous knowledge with regard to the preservation of biodiversity and its sustainable use. The Vienna Declaration and Programme of Action emerging from the 1993 World Conference on Human Rights recognizes the dignity and unique cultural contributions of indigenous peoples, and strongly reaffirms the commitment of the international community to the economic, social, and cultural well-being of indigenous peoples and their enjoyment of the fruits of sustainable development.

The draft statement of United Nations Draft Declaration on Rights of Indigenous People, expressly states that traditional lands of the indigenous people should be returned, if not they should be equally be compensated. The United Nation Forum on Forest in. the fifth session expressly stated sustainable development and management of forest would not be possible without the participation of indigenous community residing in the forest. India though a signatory to many of the convention stated above, yet has failed to implement the provisions stated in hem through relevant legislation.

3

Invertebrate Species in Forest Ecosystems

The diversity and abundance of invertebrate species in forest ecosystems are determined by the interaction of complex ecological and environmental factors. Research into these factors is gradually providing data that can be incorporated into models, both descriptive and mathematical, of the processes involved. At the core of such models is the description of the availability of resources that enables a particular invertebrate species to reproduce, whereas the rate of reproduction is dependent on multiple factors, including temperature, competition, etc., that determine utilization of available resources. In relation to exploitation of resources within a forest ecosystem, the primary focus for forest managers is the potential of invertebrates to cause damage and to reduce yields of the trees themselves. This is certainly the case in managed forests where the primary aim is to grow trees for commercial gain, often with associated value from recreational, landscape or environmental considerations. Any invertebrates that feed directly on the trees, or have an influence of how the trees grow, may be classified as pests and hence may require some form of pest management to reduce the threat. But it is pertinent to consider how ecological theory and practice can be used both to understand why a pest problem has arisen and to provide guidance on how to combat that problem. The same principles can be applied to understanding and enhancing the biodiversity of forests, including conservation of rare and beneficial invertebrates. Indeed, the two extremes of pest outbreak and invertebrate rarity are part of the same continuum of interaction between invertebrates and their host trees.

In general, considering that trees are known to harbour the greatest numbers of insect species (diversity), it is surprising that serious pest status is reached only relatively infrequently. For example, although oak *(Quercus* spp.) is known to support at least 450 species of insect in Britain, Bevan (1987) lists only 35 species as pests, of which five were classified as important or severe (XXX-XXXXX on Bevan's scale). Wider listing of pest species has been noted for forest insects in continental Europe, although even here the number of consistently serious pests is small relative to the total diversity of insects present (Klimetzek 1993). Naturally, the severity of a pest outbreak will be classified differently depending on the purposes of the forest manager, and thus it is not always possible to classify insects consistently as pests across geographical ranges or, indeed, between years. Further, an insect may normally be classified as uncommon or rare but, given an abundance of resources, may be described later as a pest. This was certainly the case for oak pinhole borer, *Platypus cylindrus* (Coleoptera: Platypodidae), which prior to the 1987 great storm in southern England was listed in the British Red Data Book as rare.

Winter (1988) warned that the great increase in availability of freshly dead oak following the storm could result in an increase in numbers of *P. cylindrus* and, subsequently, it has been noted that the insect is now causing damage to high-value oak logs as a result of its habit of boring directly into the heartwood. Therefore, consideration of why invertebrate abundance might increase above a given threshold level can provide a more consistent appraisal of status and also potential approaches for reducing populations below that threshold or indicating measures that can be taken to encourage rare or endangered species.

Among the many interacting processes that determine insect diversity and abundance, a few attributes that are particularly well linked to the environmental status of trees can be distinguished. These provide a means to describe intrinsic processes but also allow comparison with other plants, thus offering the potential for a more consistent approach to descriptive population ecology of invertebrates irrespective of the host plant being considered.

Longevity

Trees are the longest-lived components of an ecosystem and, in the majority of natural situations, approach or represent the climax stage of vegetation succession. This provides opportunities for continuity of insect population growth and over time this will give a particular set of insect associates for a given tree species. Continuing the theme of resource availability, the fact that trees may be present in the same location for tens or hundreds of years ensures that both migrant and, particularly, relatively sedentary invertebrates can exploit them at some stage during their growth. However, longevity *per se* is not a quantifiable attribute but does provide the means for the other components, to influence invertebrate diversity and abundance. Increasing stability in the environment, as represented under forest conditions, leads to stability in both microclimate and resource availability, particularly favouring exploitation by specialist insect feeders and their associates.

Biogeography

Tree species richness in a given region or locality varies considerably with geographical location and thus the climax community structure and longevity of individual species may vary accordingly. Although subject to uncertainty and lack of unequivocal evidence of the underlying processes, there is certainly a gradient of species richness of plant cover, especially trees, from the tropics to the poles. Not surprisingly, the same applies to a wide range of animals and plants; as an extreme example, the species richness of orchids ranges from 15 to 2500 in a latitudinal range from 0 to 66° S. Further examples are provided by Brown (1988). More specifically, in an analysis of species richness of deciduous forests both in the tropics and, latitudinally, within North America, Brown and Gibson (1983) contrasted the high diversity of tree species in the tropics (up to 100 ha^{-1}) with the low diversity from north to south within North America (1–30ha^{-1}). However, even within such a generalization there are exceptions, so that for example conifers are more diverse in temperate latitudes.

Geological Time

The ecological structure of forests and of their associated invertebrate faunas is determined by both present and recent characteristics, particularly area, and by their history in geological time. The latter characteristic is driven as much by *time per se* as by the history of major disturbances (glaciations, volcanic eruptions, land shifts, etc.) within geological time. Thus, present flora and fauna may well be determined by recolonization since the last major disturbance, the diversity of species being a reflection of both colonization and speciation relative to more recent ecological conditions. The gradient of species from polar to tropical regions could therefore be explained by the high level of disturbance closer to the polar masses compared with the relative stability of climate in the tropics. Evidence from habitats that have become isolated since the last major disturbance indicates that current species diversity may reflect the historical link to the original major habitat. For example, islands that were once joined by land bridges to the larger land masses of New Guinea have species diversities of birds greater than expected compared with other islands that were never connected. However, although local ecological communities can be explained by historical patterns, the gradients over wide geographical scales are not so easily attributed to time as the major determinant.

The concept that major disturbances drive change undoubtedly carries weight, but closer analysis of tropical rain forests reveals that they too have been subject to disturbances of various sorts and are, in many cases, much changed over geological time. Based on this knowledge Haffer proposed a theory that disturbances in the tropics during the Pleistocene (alternating wet and dry periods giving rise to forest blocks surrounded by grassland) gave rise to refugia in which high levels of speciation took place, ultimately leading to the current high diversity characteristic of the tropics. While this theory is plausible and has been invoked by other authors, tests of the hypothesis have cast doubt over its wide applicability.

Irrespective of the mechanistic bases for variation in species richness over wide latitudinal ranges, time is a significant determinant at a more local scale. Birks (1980) analysed the insect faunas of trees in Britain during the period since the last glaciation 13000 years ago and concluded that there was a significant correlation with the time that a given tree species had been continuously present in Britain during that period. Taking this further, Kennedy and Southwood (1984) included Birks' radiocarbon estimates of time in Britain as a variable in a multiple regression analysis of insects on British trees. They showed that time was a significant predictor, both in its own right and as a co-predictor with log abundance (area). This is logical considering that over time there will be a tendency for well-established tree genera to increase their abundance, despite the recent proliferation in areas arising from commercial afforestation.

Size

A consequence of basic structure and longevity is that trees tend to be the largest components of the phylloplane, thus providing a much wider number and range of ecological niches compared with herb and shrub layer plants. Such attributes enable more insect species to colonize the plants without facing competition, both interspecific and intraspecific. Size has many interrelated characteristics that help to explain the presence of both herbivores and their natural enemies.

- A larger physical presence is more easily detected and colonized by invertebrates. This concept, termed 'apparency', reflects a higher visual profile for colonizing adults or passively migrating immature life stages (*e.g.* mean height was a good predictor in models of the distribution of Scolytidae in Finland; Heliovaara & Vaisanen 1995) and a greater production of olfactory cues.
- Complex plant architecture is a consequence of greater size and for trees is particularly linked to the provision of many niches capable of supporting invertebrates with a wide range of feeding strategies (guilds). Although the concept of the guild is not always clearly defined, it is nevertheless a useful means of describing the trophic interactions of invertebrates and trees without having to consider their precise taxonomic status. Thus, seed, bud, leaf, bark and wood feeders can be distinguished, each occupying a different part of the tree and therefore potentially avoiding competition.
- Increased biomass provides greater food resources per individual, thus encouraging growth and breeding success whilst reducing potential intraspecific competition. Price (1992) has reviewed the role of plant resources in insect population dynamics and concludes that, in part, the carrying capacity of an ecosystem depends on plant succession

in time and space. Carrying capacity is therefore greater for trees, which generally appear late in succession, compared with herbs or shrubs. Thus biomass is available for longer and in greater quantities on trees compared with other plants.

- The concept of enemy-free space applies to the likelihood of herbivores being overlooked by searching parasitoids or predators. One aspect of size is that there is more opportunity for herbivores to occupy a part of the plant that offers protection against natural enemies (and competition from other herbivores). The complexity and number of niches present on a plant (plant architecture; Lawton 1983) therefore has an effect on enemy-free space. The greater the architectural complexity, the greater the time that natural enemies may have to spend in searching for prey, thus reducing their overall effectiveness and enhancing the survival probability of herbivores.

Area

Forests tend to occupy greater areas than other plants. There is evidence to support the view that the greater the area occupied by the host plant, the greater the number of insect species associated with that host. This is called the species-area relationship and was originally linked to the island biogeography theories of species isolation and colonization.

Species-area relationships have been demonstrated for many plants and animals, including both highly mobile and sedentary species. With regard to the numbers of insect species on trees, data on the British insect fauna, one of the best characterized in the world, have been analysed by a number of authors.

Arising from earlier tentative analyses, Kennedy and Southwood (1984) analysed the relationships between insect species and characteristics of tree species in Britain. They concluded that there was a significant relationship between the area occupied by British tree species and the numbers of insect species associated with them. The area occupancy data were based on county and subcounty information of plant records and were thus potentially overestimates of actual areas occupied.

Claridge and Evans (1990) reanalysed the species-area relationships using the same insect data as Kennedy and Southwood but employing actual area information derived from Forestry Commission census data. They concluded that area was not a good descriptor for insect species on deciduous trees but was just significant for conifers.

However, further unpublished analysis of the data broken down into numbers and areas of habitat patches occupied by the tree genera indicated that whereas total area was not a good descriptor, the degree of fragmentation of that area provided a better explanation for the observed insect species numbers. A similar conclusion, concentrating on Lepidoptera that have moved hosts to include conifers, was reached by Fraser and Lawton (1994) in an analysis of new associations on British trees. The overall finding that area provides a broad descriptor of insect diversity remains valid, although this does not explain the mechanisms that lead to these associations.

Taxonomic Relatedness

In a given ecosystem, the closer the taxonomic relatedness of host plants, the more similar the numbers of insect species associated with those plants. This has been documented for tree species that have a long history in a particular location and thus provide rich sources of invertebrate species that may subsequently colonize other tree species, especially if they are taxonomically related. A particularly good example of the process in action is the colonization of *Nothofagus,* a member of the Fagaceae newly introduced to Britain, which has been colonized rapidly by insects associated with closely related members of the native Fagaceae.

Geographical isolation as a critical factor in determining invertebrate diversity and implications for international movement of pest organisms. Geographic isolation drives the acceleration of speciation and provides the means to derive principles on the processes involved. In addition, the fact of isolation means that some invertebrates abundant in one region will be absent elsewhere, despite conditions being suitable for their establishment. In such situations, it is pertinent to examine the international plant health implications of geographical isolation and how modern patterns of trade influence the degree of isolation and later colonization by new insect species.

Geographical Isolation as a Factor in Determining Insect Diversity

In general, the basic principles apply across ecosystem boundaries and geographical zones extending from the poles to the tropics. In addition, the isolation afforded by relatively small islands provides further evidence of the principles of species colonization over space and time. Great Britain provides a particularly good example, because of its island position in the Atlantic Offshore climatic zone and the fact that it has been subjected to glaciation in recent geological time.

Among the ecological characteristics that arise from this set of circumstances are a relatively depauperate mix of both animal and plant species that have arisen from the combination of losses resulting from the last ice age and the slow rate of recolonization arising from island status. British flora and fauna are therefore impoverished relative to continental Europe. Further influences arise from the preponderance of exotic tree species in commercial forestry in Britain, which although dominated by exotic conifer species also has increasing numbers of exotic broadleaved species.

Expanding on the principle that colonization of tree species is determined by the features, the ranges of numbers of insect species associated with trees in Britain reflect both the area (with provisos) and length of time that trees have been continuously present in Britain. However, there are exceptions, so that Sitka spruce in particular would appear to support far fewer species than predicted on the basis of area, although this probably also reflects the relatively short history of the species and recent expansions in areas planted in Britain.

Implications for International Movement of Pest Insects

Trading patterns between countries have had a considerable influence on both the flora and fauna of most of the countries in the world. This reflects deliberate introductions of plants and animals for agricultural, forestry and amenity purposes and also accidental introductions by association with other commodities. In the majority of cases, the introduced species have had a neutral to beneficial effect, reflecting their 'domesticity' within managed crop or animal husbandry systems. However, there are many exceptions to this generalization and in most cases the results of introduction of new species were not foreseen, either through lack of knowledge of the potential impacts or because the introduction was accidental and undetected. The literature on international biological control is dominated by examples where attempts have been made to rectify the effects of new pest introductions (both animal and plant). A classic example is the control of the woodwasp *Sirex noctilio* (Hymenoptera: Siricidae) in exotic pine forests in Australia with the introduced nematode *Deladenus siricidicola* (Nematoda: Neotylenchidae) isolated from *Sirex* in New Zealand. This nematode has a life cycle that includes a free-living form that feeds on the fungus *Amylostereum areolatum* and a parasitic form that invades *S. noctilio* larvae and subsequently migrates to the reproductive organs of the female wasp, causing sterility. In this situation, all elements of the interaction are exotic to the new location, with little prospect of natural control being realized.

ASCOMYCOTA

Mycosphaerella pini Rostrup

Mycosphaerella pini is a fungus that infects and kills the needles of *Pinus* spp. resulting in significant defoliation, stunted growth and eventually death of host trees although susceptibility among pine species does vary. Native to temperate forests of the northern hemisphere, it is a major pest of pine plantations in the Southern Hemisphere, where both the host and the pathogen have been introduced. The widely planted *P. radiata* is particularly susceptible and many forests planted with this species in the Southern Hemisphere, particularly in East Africa, New Zealand and Chile, have been devastated by this needle blight. This pathogen has forced managers in some areas to abandon the planting of *P. radiata* and depend more on other tree species. In its native range *M. pini* normally causes little damage, but since the late 1990s it has been causing extensive defoliation and mortality in young plantations of lodgepole pine (*Pinus contorta* var. *latifolia*) in northwestern British Columbia, Canada (Woods, Coates and Hamann, 2005). Mortality of mature lodgepole pines has been observed in mixed-species stands, where scattered pine represents only a small proportion of stand composition; this represents a globally unprecedented occurrence for *M. pini* (Woods, Coates and Hamann, 2005). The current epidemic coincides with a prolonged period of increased frequency of warm rain events throughout the mid-to-late 1990s allowing for the rapid spread and increased rates of infection. Unlike many other pests, changes in precipitation patterns may be more important than changes in temperature for predicting the spread and impact of *M. pini*.

BASIDIOMYCOTA

Armillaria Species

Armillaria species are common worldwide pathogens of trees, woody shrubs and herbaceous plants that can cause wood decay, growth reduction and even mortality, particularly in trees stressed by other factors, or in young trees planted on sites from which infected hosts have been removed. *Armillaria* species can become more aggressive and damaging when elevated temperatures cause drought stress thereby reducing tree defences. Tree physiological condition in general may be an important factor in controlling the impacts of *Armillaria* species, and climate change may affect their epidemiology.

COLEOPTERA

Agrilus Pannonicus

A number of Buprestid beetles of the genus *Agrilus* have been linked to oak decline. Incidences of these species have increased worldwide (both in their countries of origin and by international movement) and their impacts are being linked to host tree stress potentially caused by climate change. For example, *Agrilus pannonicus* (=*A. biguttatus* (Fabricius)) has recently been associated with a European oak decline throughout its natural range and has increased in incidence in several countries including France, Germany, Hungary, Poland and the Netherlands, and the UK where it is believed to be contributing to oak decline. Infestations can result in extensive tree mortality which, combined with other factors involved in the decline, can drastically alter the species composition of oak forests.

Dendroctonus Frontalis Zimmermann

Dendroctonus frontalis is considered to be one of the most damaging species of bark beetles in Central America and southern areas of North America. It is a major pest of pines and has a wide distribution occurring from Pennsylvania in the United States south to Mexico and Central America. Populations can build rapidly to outbreak proportions and large numbers of trees are killed. Initial attacks are generally on weakened trees however *D. frontalis* is capable of killing otherwise healthy trees. This beetle kills trees by a combination of two factors: girdling during construction of egg galleries; and the introduction of blue stain fungi of the genus *Ophiostoma.* Because of their short generation time, high dispersal abilities and broad distribution of suitable host trees, the southern pine beetle has the potential to respond quickly and dramatically to any changes in climate.

In October 1998, Hurricane Mitch hit Central America, causing floods and mudslides that ravaged local communities, forests and infrastructure. In the years that followed an unprecedented regionwide outbreak of pine bark beetles, mainly *D. frontalis* in association with other *Dendroctonus* and *Ips* species, destroyed over 100 000 ha of pine forest. As most of the standing dead and felled trees were left on site, fuel loads were drastically increased thereby resulting in extensive wildfires. With climate change expected to increase the frequency and severity of extreme events such as hurricanes, the potential for future devastating impacts on forests from both the initial disturbance and its cascading effects (*i.e.* other disturbances such as pest outbreaks and fire) is quite high. Warmer temperatures attributed to climate change have also influenced the southern pine beetle resulting in range expansions in the United States. Laboratory measurements and published records of mortality in wild populations indicate that a temperature of -16 °C or less result in almost 100 percent mortality of the pest, thereby limiting its distribution in its current northern range. It was predicted that an increase in temperature of 3 °C would enable outbreaks to occur approximately 178 km farther north than in historical times (Ungerer, Ayres and Lombardero, 1999). Recent outbreaks of the southern pine beetle in northern and high-altitude ecosystems, where they were previously rare or absent, have been attributed to a warming trend of 3.3 °C in minimum winter air temperatures in the southeastern US from 1960-2004. This northern expansion is about as predicted by Ungerer, Ayres and Lombardero (1999). The southern pine beetle has also possibly adapted ways to increase survival in cooler climes. Tran *et al.* (2007) showed, through field and laboratory studies of a northern population, that prepupae were more cold tolerant (by more than 3 °C) than pupae, adults and feeding larvae, and that the winter life stage structure was strongly biased toward this most cold-tolerant life stage. This tendency to overwinter in a cold tolerant life stage could be a coincidence however rather than a true adaptation.

Dendroctonus Ponderosae

The mountain pine beetle (*Dendroctonus ponderosae*) is the most destructive pest of mature pines in North America, particularly lodgepole pine

(*Pinus contorta*). In the western United States, outbreaks have been increasing in area after several years of drought (Tkacz, Moody and Villa Castillo, 2007). A major epidemic of this pest has also been ongoing in western Canada (British Columbia (BC), and more recently, Alberta) for several years and even with large-scale efforts to mitigate the impacts of the pest, millions of trees have been killed. A record of over 10 million hectares of pines were recorded as infested during 2007 aerial overview surveys in BC, with 860 973 ha of this located in provincial parks and protected areas. It has been predicted that if the beetle continues to spread at its current rate as much as 80 per cent of mature pine in BC will be dead by 2013. The large numbers of dead and dying trees have also increased the risk of wildfires.

The problem has been exacerbated by successive years of mild winters, resulting in decreases of mortality of overwintering stages and generation time. Their life cycle is generally completed in one year; warmer temperatures can result in two generations per year while cooler ones may results in one generation every two years (Amman, McGregor and Dolph, 1990). Drought conditions associated with warmer temperatures have also weakened the trees and increased their susceptibility to the beetles. Warmer temperatures have thus opened up previously climatically unsuitable mature pine stands to the pest.

Dendroctonus Rufipennis

Dendroctonus rufipennis is a North American pest of spruce, particularly white spruce (*Picea glauca*) and black spruce (*P. mariana*) in the north, Engelmann spruce (*P. engelmannii*) and sitka spruce (*P. sitchensis*) in the west, and red spruce (*P. rubens*) in the east. It tends to attack weakened or windthrown trees and outbreaks are mostly linked to predisposing factors. As a result it can be expected that the impacts of climate change on trees and forests could enhance spruce beetle outbreaks. In fact, Hebertson and Jenkins (2008) investigated the impact of climate on spruce beetle outbreaks in Utah and Colorado, USA between 1905 and 1996 and found that historic outbreak years in the intermountain region were related to generally warm fall and winter temperatures and drought conditions. Similarly outbreaks in both Canada (Yukon Territory) and the US (Alaska) appear to be related to extremely high summer temperatures which influenced spruce beetle population size through a combination of increased overwinter survival, a halving of the maturation time from two years to one year, and regional drought-induced stress of mature host trees.

HEMIPTERA

Aphids (Aphididae)

With short generation times and low developmental threshold temperatures, aphids are a group of insects that can be expected to be strongly influenced by environmental and climatic changes. In general, it has been predicted that aphids will appear at least eight days earlier in the spring within 50 years, though the rate of advance will vary depending on location and species. This could potentially result in greater damage to host plants depending on the phenology of host plants and natural enemies.

Zhou *et al.* (1995), for example, investigated the timing of migration in Great Britain for five aphid species (*Brachycaudus helichrysi, Elatobium abietinum, Metopolophium dirhodum, Myzus persicae*, *Sitobion avenae*) over a period of almost 30 years and concluded that temperature, especially winter temperature, is the dominant factor affecting aphid phenology for all species. They found that a one degree Celsius increase in average winter temperature advanced the migration phenology by 4-19 days depending on species.

Elatobium abietinum (Walker) (Aphididae)

The green spruce aphid (*Elatobium abietinum*) is also believed likely to benefit from the increase in winter survival, leading to more intense and frequent defoliation of host spruce trees (*Picea* spp.). This aphid is native to Europe but has also been reported in both North and South America.

Infestations in Great Britain have resulted in large losses of spruce foliage and height both during the active infestation and in subsequent years. Westgarth-Smith *et al.* (2007) showed that warm weather associated with a positive North Atlantic Oscillation (NAO) index caused spring migration of *E. abietinum* to start earlier, last longer and contain more aphids. Positive NAO values correspond to warmer atmospheric conditions over Great Britain. Since global warming is believed to increase NAO variability, shifting the system to more positive values, this will most likely lead to further increases in aphid activity and more damage to spruce trees and forests in the area.

HYMENOPTERA

Cephalcia arvensis Panzer

The spruce webspinning sawfly is monophagous on spruce (*Picea*) and endemic to the spruce range in Eurasia, where outbreaks have been seldom recorded. From 1985-1992 however there was a sudden outbreak of the sawfly in the Southern Alps during which populations developed an annual life cycle and grew exponentially, causing repeated defoliations resulting in extensive tree death. *Cephalcia* species generally show low fecundity and have an extended diapause of a few years that is stimulated by low temperatures at pupation time. The outbreak corresponded to a period of high temperatures and low precipitation and severe water stress for the host trees. As a result, the insect was able to adapt to the new climate resulting in lower mortality, faster development and higher feeding rates of the sawfly. In addition, the sudden increase in population density was not quickly followed by that of natural enemies, thereby allowing for unlimited population growth.

Neodiprion Sertifer

The European pine sawfly *Neodiprion sertifer* is an important pest species on pines in Europe, northern Asia, Japan and North America where it was introduced. It is one of the most serious defoliators of Scots pine (*Pinus sylvestris*) forests in northern Europe. Virtanen *et al.* (1996) suggested that outbreaks of the sawfly on Scots pine in eastern and northern Finland are prevented by low winter temperatures which kill eggs, and predicted that outbreaks would become more common with winter warming. A high variation in freezing avoidance of eggs was also noted which would allow *N. sertifer* to adapt to predicted climate change and spread its distribution northwards.

LEPIDOPTERA

Warmer temperatures have been linked to increasing populations of forest Lepidoptera species.

Butterflies

While only a few species of butterflies are considered to be serious forest pests, some of the best, and most researched, examples of the impacts of climate change on insect distributions and phenology have been butterflies. The geographic ranges of many species have shifted northward and upwards in elevation associated with climate warming, leading to increases in species richness at high latitudes and elevations and in some cases possible local extinction at lower altitudes.

Range expansions in butterflies have been well documented and changes in butterfly phenology have also been reported. In the UK, species have been advancing their flight periods by approximately 2-10 days for every 1 °C increase in temperature. Similar changes in phenology as a response to warming has been noted in Spain where butterflies have advanced their first appearance by 1-7 weeks in 15 years and in California, USA which has seen an advancement of approximately eight days per decade.

Some species-specific examples of the influence of climate change include the following.

- The African Monarch butterfly (*Danaus chrysippus*) has spread northward, establishing its first population in southern Spain in 1980 followed by the establishment of multiple populations along the east coast of Spain.
- Warmer temperatures have increased survival and facilitated a latitudinal and altitudinal range expansion of *Atalopedes campestris* in the western USA.

- Edith's checkerspot butterfly (*Euphydryas editha*) has shifted its distribution northwards and also upwards in altitude in North America. Populations at the northern edge of the species range in Canada and also at higher altitudes within the main range have experienced increased survival whereas populations at the southern edge in Mexico have declined.
- In Europe the speckled wood butterfly (*Parage aegeriae*) has increased its range northward beyond its original, primary host.
- The black-veined white butterfly, *Aporia crataegi*, has expanded its altitudinal range in the mountains of the Sierra de Guadarrama of central Spain resulting in local population extinctions at lower warmer altitudes. While climate is becoming less of a limiting factor in its distribution at higher altitudes, it is however limited by the absence of host plants.

Choristoneura Fumiferana

The spruce budworm, *Choristoneura fumiferana*, is a major defoliator of coniferous forests across North America. Balsam fir (*Abies balsamea*) is the preferred host but they readily attack white, red and black spruce (*Picea glauca*, *P. rubens*, *P. mariana* respectively) and may even be feed on tamarack (*Larix* spp.) and hemlock (*Tsuga* spp.). Outbreaks of this budworm can persist for 5-15 years with periods of 20-60 years in between. In eastern Canada the period of population cycle has averaged 35 years over the last 270 years. During uncontrolled outbreaks they can kill almost all trees in dense, mature stands of fir. Climatic influences on life history traits are considered a major factor in restricting outbreaks and as a result, a changing climate is expected to impact the severity, frequency, and spatial distribution of spruce budworm outbreaks. The success of the insects in establishing feeding sites in the spring depends on initial egg weights and synchrony of their development with that of buds of their hosts which is strongly influenced by climatic factors. This synchronization is critical in initiating outbreaks and thus determining the intensity of damage. However the spruce budworm is able to tolerate some asynchrony between spring emergence and vegetative shoot development as the second instars have adapted morphologically and behaviourally allowing them to mine needles.

In parts of its range, particularly at northern extremes, temperature can also influence the duration of outbreaks as collapses are often associated with the loss of suitable foliage often as a result of late spring frosts. Normal collapse of the outbreak in the core range of host trees is associated with mortality caused by natural enemies late in the larval stage. Natural enemies of the spruce budworm, *C. fumiferana*, are less effective at higher temperatures and therefore climatic factors have the potential to enable massive outbreaks of this pest providing there is suitable availability of host trees.

Epirrita Autumnata

Epirrita autumnata is a holarctic species that has been expanding its outbreak range in some areas. In the Nordic countries of Europe, *Epirrita autumnata* outbreak cycles are typically most prevalent in northernmost and continental birch forests but during the past 15-20 years it has expanded into the coldest, most continental areas previously protected by extreme winter temperatures. This pest overwinters in the egg stage and therefore the level of egg survival is dictated by minimum winter temperatures. Virtanen, Neuvonen and Nikula (1998) investigated the relationship between *E. autumnata* egg survival and minimum winter temperatures in northernmost Finland. They predicted that climate warming would result in a two-third reduction of the area of forests with winter temperatures cold enough to keep *E. autumnata* populations low by the middle of the next century. A rise in winter temperatures therefore will likely increase the area of forest susceptible to damage by the autumnal moth.

Lymantria Dispar Linnaeus

The gypsy moth, *Lymantria dispar*, is a significant defoliator of a wide range of broadleaf and even conifer trees. While low population levels can exist for many years without causing significant damage, severe outbreaks can occur resulting in severe defoliation, growth loss, dieback and sometimes tree mortality. Two strains of gypsy moth exist – the Asian strain, of which the female is capable of flight; and the European strain, of which the female is flightless. The Asian strain is native to southern Europe, northern Africa, central and southern Asia, and Japan and has been introduced into Germany and other European countries where it readily hybridizes with the European strain. It has also been introduced but has not established in Canada, the US and the UK (London). The European strain is found in temperate forests throughout Western Europe and has been introduced into Canada and the US. The gypsy moth is considered a significant pest in both its native and introduced ranges.

There has also been a noted increase in outbreaks in areas previously unaffected by this pest such as the Channel Islands (Jersey) and new areas in the UK (Aylesbury, Buckinghamshire). In Canada the spread of the gypsy moth has so far been prevented by climatic barriers and host plant availability as well as by aggressive eradication of incipient populations. However it is predicted that the gypsy moth will be able to extend its range in North America as a result of higher overwinter survival of egg stages because of milder winters and higher accumulation of day degrees for larval development. Similar predictions have been made for other areas. For example, Pitt, Régnière and Worner (2007) noted an increase in the probability of establishment of the gypsy moth in New Zealand, particularly in the South Island.

Increasing atmospheric concentrations of CO_2 may also influence the severity of gypsy moth outbreaks. Larval performance on host plants grown under elevated CO_2 varies depending on the species, being reduced on some hosts such as aspen and increased on others, such as oak.

Lymantria Monacha

Lymantria monacha is a major pest of broadleaved and coniferous trees in Europe and Asia. Defoliation by nun moth larvae can kill host trees especially conifers and has caused extensive losses despite intervention with biological and chemical insecticides. In parts of Europe, the occurrence of outbreaks has increased possibly as a result of the establishment of extensive pine plantations in poor quality areas or as a result of a changing climate. It has been predicted that nun moth will spread northwards in Europe because of higher accumulated day degrees and improved overwinter survival. Using modelling software, Vanhanen *et al.* (2007) predicted that climate warming would shift the northern boundary of distribution north by approximately 500-700 km and the southern edge of the range would retract northwards by 100-900 km.

Operophtera brumata Linnaeus

Operophtera brumata is distributed throughout Europe, North Africa, Japan and Siberia and has also been introduced into Canada and the US. It feeds on a variety of deciduous trees and shrubs including apricot, cherry, apple, plum, blueberry, crab-apple, sweet chestnut, red currant and black currant, oaks, maples, basswood and white elm. Climate change is impacting the spread of the winter moth. In the Nordic countries of Europe, Jepsen *et al.* (2008) noted that *O. brumata* had been climatically restricted to more southern and near-coastal locations in the regions but warmer temperatures has resulted in expansions in its outbreak area further northeast. While increased temperatures appear to assist the winter moth expand its distribution, it appears that they do not have the same impact on its natural enemies which may allow populations of this pest to grow unchecked.

Climate change has affected the phenology of many species in different ways. In the Netherlands over the past 25 years, early spring temperatures have increased while winter temperatures have not. As a result a climate change induced asynchrony has occurred between winter moths and their host, pedunculate oak, *Quercus robur*, with eggs hatching before bud burst. Such a situation leaves no food for the larvae resulting in starvation and death. This also has implications for other species that depend on the larvae for food such as the great tit (*Parus major*) which feed *O. brumata* caterpillars to their young. While both egg hatch and bud burst have advanced over the last 25 years, egg hatch has advanced much more leading to a decrease in synchrony from a few days to almost 2 weeks. However others have noted that, while warmer temperatures have lead to earlier egg hatch, the autumnal pupal diapause of the winter moth is prolonged at higher temperatures thereby counteracting the impact and resulting in a life cycle that is not shortened overall. Differences between observations of synchronicity between moth egg hatch and host bud burst may result from regional, intra-specific differences.

Thaumetopoea pityocampa

The pine processionary caterpillar, *Thaumetopoea pityocampa*, is considered one of the most important pests of pine forests in the Mediterranean region (EPPO/CABI, 1997). It is a tent-making oligophagous caterpillar that feeds gregariously and defoliates various species of pine and cedar. The life cycle of the pine processionary caterpillar is typically annual but may extend over two years at high altitudes or in northern latitudes. At northern latitudes and at higher altitudes, adults emerge earlier. Climate change is having clear impacts on the distribution of this important forest pest. Battisti *et al.* (2005) reported a latitudinal expansion in north-central France of 87 km northwards from 1972-2004 and an altitudinal shift of 110-230 m upwards in the Alps of northern Italy from 1975-2004 and attributed the expansions to reduced frequency of late frosts, which increases survival of overwintering larvae, as a result of a warming trend over the past three decades. In the last ten years the pine processionary caterpillar has spread almost 56 km northward in France. During the summer of 2003, the warmest summer in Europe in the last 500 years, *T. pityocampa* exhibited an unprecedented expansion to high elevation pine stands in the Italian Alps, increasing its altitudinal range limit by one third of the total altitudinal expansion over the previous three decades. This unusual and fast spread has been attributed to increased nocturnal dispersal of females during the unusually warm night temperatures. The gradual warming of the region has allowed the pest to maintain its presence at this altitude because of increased larval survival.

In the Sierra Nevada mountains of southeastern Spain, *T. pityocampa* has expanded to higher elevations over the last 20 years as a result of increasing mean temperature.

Relict populations of Scots pine (*Pinus sylvestris* var. *nevadensis*) occurring within this newly expanded range of the caterpillar are being increasingly attacked, particularly in warmer years. This range expansion caused by climate change has potentially devastating consequences for this endemic mountain species which is likely to suffer from the direct effects of climate change as well. Given that the present distribution of the *T. pityocampa* is not constrained by the distribution of its hosts, that warmer winters will increase winter larval feeding activity, that the probability of lower lethal temperature will decrease, it can be expected that the improved survival and spread into previously hostile environments will continue.

Thaumetopoea Processionea

Native to central and southern Europe, *Thaumetopoea processionea* is a major defoliating pest of oak. Since the late 20th century it has been expanding its range northwards and is now firmly established in Belgium, Denmark, northern France, and the Netherlands and has been reported from southern Sweden and the UK. It is believed that the northward progression of the oak processionary moth is due to improved synchrony of egg hatch and reduction of late frosts as a result of warmer temperatures.

Zeiraphera Diniana Guenée

The larch bud moth, *Zeiraphera diniana*, is a European pest that has been defoliating large areas of larch forests in the Alps every 8-10 years for centuries. It has an annual life cycle, overwintering as an egg on the larch branches and feeding on the needles as soon as the bud breaks. As such synchrony between egg hatch and bud burst is critical. Increased temperatures associated with climate change have affected this relationship leading to asynchrony and reduced incidences of the moth in Switzerland. It has been reported that abnormally high temperatures result in unusually high egg mortality.

PHYTOPHTHORA RAMORUM WERRES

Phytophthora ramorum causes a very serious disease called sudden oak death which causes extensive mortality of tanoak and oaks. It is also associated with disease on ornamental plants and other broadleaf and conifer trees. This pathogen is a significant problem in both North American and European forests and nurseries. The geographic origin of *P. ramorum* is unknown; it is believed that it has been introduced independently to Europe and North America from an unidentified third country. The pathogen likely disperses through a variety of means. Sporangia may be dispersed locally by rain splash, wind-driven rain, irrigation or ground water, soil and soil litter. Bark and ambrosia beetles are commonly found on infected trees but their potential role of vectors has not yet been investigated. Consequently, changes in climate, precipitation and temperature in particular, will likely produce more optimal conditions for the pathogen resulting in an increase in disease occurrence.

NEMATODES

In general there is a close correlation between soil temperatures and the distributions of some plant-parasitic species of nematode. For example, *Meloidogyne incognita*, previously deemed limited to the Mediterranean area, was recently found in the Netherlands. It is also believed that a one degree Celsius rise in temperature would allow *Longidorus caespiticola* to become established further north in Great Britain.

Bursaphelenchus Xylophilus

The pine wilt nematode, *Bursaphelenchus xylophilus*, is the causal agent of pine wilt disease and is spread by *Monochamus* beetles. Native to North America where it is not considered a serious pest, the nematode is a major threat to Asian and European pine forests and has resulted in extensive tree mortality in countries where it has been introduced. Changes in both temperature and precipitation are likely to impact the spread of the nematode and the severity of damage caused by the disease. Pine wilt disease is most prevalent in warm climates as the nematode completes its life cycle in 12, 6 and 3 days at 15, 20 and 30°C, respectively. High temperatures and low precipitation in summer cause accelerated damage through their impacts on vector activity, propagation of the nematode and water stress on trees. In Japan, while annual tree losses to the disease have gradually decreased, infestations have spread into northern areas and into forests at higher elevations as a result of increased temperatures.

OOMYCOTA

Phytophthora Cinnamomi

Phytophthora cinnamomi is considered one of the most widely distributed and destructive forest pathogens. It has a wide host range infesting over 1000 species resulting in root rot and cankering. The native range is unknown but it is believed to be from Southeast Asia and southern Africa. Currently the pathogen can be found in most temperate and subtropical areas in the world in Africa, Asia and the Pacific, Europe, Latin America and the Caribbean, Near East, and North America. In most countries it is only known in nurseries but in Europe (France, Italy, Spain, Portugal) it has observed in natural environments. Temperature, moisture and pH all influence the growth and reproduction of the fungus. In a study on the impacts of climate warming on *P. cinnamomi*, Bergot *et al.* (2004) predicted a potential range expansion of the disease in Europe of one to a few hundred kilometres eastward from the Atlantic coast within one century.

ECOSYSTEM IN SPECIES OF THE GREATER HIMALAYA

Two subspecies of goral occur in India, *viz.*, the Grey goral (N.g.goral) distributed in the Western Himalaya (upto 3,300m) and Shiwaliks in the states of Jammu & Kashmir, Himachal Pradesh, Uttaranchal and Haryana; and the Brown goral (N.g.hodgsni) in the Eastern Himalayas and the hills of N.E. India in the states of Sikkim, northern West Bengal, Arunachal Pradesh, Nagaland, Meghalaya and Mizoram. Another species, the Red goral (Nemorhaedus baileyi) is reported to occur in a small region in Arunachal Pradesh at the junction of Tibet, Yunnan and Myanmar. Inhabits sparsely wooded slopes with open grassy patches in a wide range of habitats that includes alpine meadows, subalpine forests, temperate forests, subtropical pine forests, tropical moist deciduous forests of Shiwaliks, and montane wet temperate and evergreen forests in N.E. India.

A monomorphic mountain goat as body size, horn size and colouration are same in both sexes. However, adult males could be distinguished from females based on the body size and horn size, if observed at close quarters. The general body colour of Grey goral is yellowish grey suffused with black and the hair on the pale area is not tinged with rufous or brown. The chin, upper lip, undersides of jaws and throat patch are white. The dark spinal stripe, if present, does not pass beyond the withers. There is no stripe down the middle of the tail and none up the back of the thighs. The body colour of Brown goral is golden or rufous brown specked with black. The black spinal stripe reaches to the root of the tail. A dark ill-defined stripe runds up the back of the thighs from the hocks. Both males and females have short insignificant sharp horns black in colour which diverge slightly, curve backwards, marked with rings or ridges.

A cliff-dwelling mountain goat occurs singly or in pairs or in small parties of 3-5 individuals. Aggregations of >5 individuals (largest reported 13 individuals) may occur during feeding depending upon the food resource availability and distribution. Predators are the Common leopard and the Himalayan yellow-throated Marten. When approached by a predator or human, goral makes an alarm call "Hiss" runs for escape cover or escape terrain, and may stop to look back. Actively feeds during early mornings and late afternoons resting under cover during noon. Goral are primarily grazers, although they feed upon tender shoots of certain shrubs and leaves of trees when available.

BHARAL, HIMALAYAN BLUE SHEEP

Pseudois nayaur, Hodgson 1833

Vernacular Names: Knabo (Ladakhi, Tibetan),nabo (Spiti), bharal (Hindi, Pahari)

Taxonomy

Family:Bovidae

Sub family: Caprinae

Tribe: Caprini

Conservation Status

IUCN: Low risk

CAMP: Low risk, least concern

IWPA: Schedule I

CITES: Not listed

US ESA: Not listed

Central Asian mountains-from Kunjereb in Pakistan in the west, to the Sichuan and Gansu provinces of China in the east, through bulk of the Tibetan plateau. Southern limit of distribution is along the Greater Himalayan chain in

India, Nepal and Bhutan. Northern flank covers the Altai Shan, and parts of the Tien Shan. In India, found in Ladakh, Lahul-Spiti, and in some high alpine meadows of Kullu, Kinnaur, Uttaranchal, Sikkim, and the western Tawang region of Arunachal Pradesh covering ca. 30,000 km^2. Most of these, apart from Ladakh and Lahul-Spiti, are narrow regions on the southern slopes of the Greater Himalaya itself.

The adult male's coat is slate grey, sometimes with a bluish sheen. The underparts and back of the legs are white, while the chest and fronts of the legs are black. Separating the grey back and white belly is a black coloured stripe. The ears are small, and the bridge of the nose is dark.

The horns are found in both sexes, and have small ridges on the upper surface. In males, they grow upwards, then turn sideways and curve backwards, looking somewhat like a moustache. Females are light brown coloured with small horns, usually diverging at the top (important feature to distinguish with ibex females in areas where they are sympatric).

Bharal live at mid to high elevations, and may climb up to the vegetation line at 5,500 m. In India, they usually occur between 2,500 and 5,500 m, but usually below this upper limit. Uses primarily moderate to steep slopes, interspersed with 'escape terrain' in the form of rocky overhangs and cliffs, where they flee to escape predators such as snow leopard and the Tibetan wolf.

Habitat similar to ibex but can occur in more open slopes and in much moister areas on the southern slopes of the Himalaya. In winter, may descend if possible or remain partial to south facing slopes and areas where snow deposition is relatively lower such as windblown ridgelines, crests and avalanche paths.

Occurs in small groups of less than 20 animals. Sexual segregation is most common in summer, when males may use more open meadows at higher elevations. Bharal are active during the day, with a bimodal peak during morning and late afternoon.

Intermediate forager, diet consists primarily of grasses, hardy herbaceous plants, also some shrubs.

FLARE HORNED OR KASHMIR MARKHOR

Capra falconeri, Wagner 1839

Vernacular Names: markhor (Punjab and Kashmir), raphoche (male) rawache (female)

Taxonomy

Family: Bovidae

Sub family: Caprinae

Tribe: Caprini

Conservation Status: Sparsely wooded mountainous regions in the western Himalayan Pir Panjal range at an elevation of 600-3,600 m. In India, they occur only in a few pockets of the Kashmir valley, notably the Kisthwar National Park, in the Greater Himalaya, with possibilities of their occurrence in the Pangi valley of Himachal Pradesh too. Bulk of the distribution is however in the Trans-Himalayan areas, now in the PoK. Other subspecies of markhor occur in the Salt, Safed Koh, Hindu Kush and Kirthar ranges in Pakistan, in eastern Afghanistan and southern Tajikistan.

The grizzled light brown to black coat is smooth and short in summer, growing longer and thicker in winter. Males have long hair on the chin, throat, chest, and shanks, while females have smaller fringes. The lower legs have a black and white pattern. Males have loosely curled, corkscrew-like horns, starting close together at the head, but spreading towards the tips. In males, they can grow up to 160 cm, and up to 25 cm in females. Female horns are also slightly twisted. A dark lateral stripe is present in males and is usually fainter in females.

Occurs in low to mid elevation, open and slightly wooded arid tracts in the Trans-Himalaya and parts of the Greater Himalaya. May migrate up to 4,000m in summer, but essentially remain in areas where rugged valleys and mountains are available at ca. 2,200m. Distinctly avoid areas with high snow cover.

Lack underwool and thus avoid excessively cold areas, but can tolerate high temperatures reaching up to 45°C. The markhor's alarm call resembles the nasal "a" not very different from the common domestic goat. The primary natural predators are leopard, Tibetan wolf, snow leopard, and lynx. Markhor, a wild goat species, are efficient negotiators of steep cliffs and use this to escape predators The markhor is mainly active in the early morning and late afternoon. An intermediate forager, it consumes primarily grasses and forbs during spring and summer months, while in the winter it turns primarily to browse for nourishment. Markhor often stand on their hind legs in order to reach high vegetation; they are known to climb the gnarled trees for foraging. During the rut, males fight for breeding rights. These competitions involve lunging and locking the horns, followed by the combatants twisting and pushing in an attempt to make the other lose his balance.

Females and young live in herds of around 9 animals; adult males are usually solitary.

HIMALAYAN TAHR

Hemitragus jemlahicus H.Smith 1826

Vernacular/other Names: jungli bakri (Hindi), thar, karth (male), tanne, bakri (female) jharal, jharal-thar, (local names) kras, jagla (Kashmiri) yang, reiwo (nepali)

Taxonomy

Family: Bovidae

Subfamily: Caprinae

Conservation Status
IUCN: Vulnerable
CAMP: Low Risk
IWPA: Schedule I
CITES: Not listed
U.S. ESA: Not listed

Inhabits a narrow strip along the southern side of the Greater Himalaya between 2,000 and 4,400m in the states of Jammu & Kashmir, Himachal Pradesh, Uttaranchal & Sikkim. Temperate and Subalpine forests interspersed with open steep slopes, temperate and subalpine scattered tree and scrub habitats, alpine scrub and meadow habitats.

A mountain goat with finely formed head, narrow erect ears, a heavy body, and long robust limbs. The body is covered with tangled masses of coarse, flowing hair, particularly well developed in adult males and forms a prominent ruff and mane that may attain a length of 25 to 30 cm, up to the knees in males. They also have a long mantle of hair that drapes from the side and rump. The fur is deep reddish brown, and there is dark mid dorsal streak, not always distinct.

Old males are darker, particularly on the back and quarters. Females and young males are lighter brown and the kids much paler. Both male and female have massive horns almost of equal size. The horns are close-set, stout at the base, keeled in front and wrinkled except towards the tips. They curve backwards, and in old males continue downwards. Group-living mountain goat that inhabits the most inaccessible terrain characterized by open steep slopes, scattered trees, scrub and grassy blanks, cliffs and rocks. Tahr live in mixed herds (1-77) for most of the year excepting for a brief period in spring when many adult males segregrate after rut to higher altitudes. Adult females and juveniles have fairly fixed home ranges and move considerably between altitudes. Predators are snow and common leopards and Himalayan yellow-throated marten. Poaching for meat by man is common. They seldom stray far away from cliffs and precipitous terrain that serves as escape terrain. Tahr are good climbers and traverse ledges and rock faces with great agility to reach scattered patches of vegetation. Tahr is a grazer feeding mainly on grasses, sedges, herbs, ferns and mosses. Prefers to feed actively during early mornings and late afternoons, and resting during mid day.

HIMALAYAN MUSK DEER

Moschus chrysogaster Hodgson 1839
Vernacular/other Names: kasturi mrig, kastura, bina,
(local Names), roos (kashmiri)
Taxonomy
Family: Moschidae

All along the southern side of the Greater Himalaya between 2,500m to 'treeline' in Jammu & Kashmir, Himachal Pradesh, Uttaranchal, Sikkim & Arunachal Pradesh. Upper Temperate, Subalpine forests, alpine scrub and meadow habitats. Two sub species of Himalayan musk deer occur in India, M.c.chrysogaster distributed in western Himalaya and M.c.leucogaster distributed in eastern Himalaya. Two other species, the Black Musk deer M.fuscus and the Dwarf Musk deer M.berevoskii are also reported to occur in some parts of Arunachal Pradesh that are close to China and Myanmar borders, but needs confirmation.

Not a true deer, but a primitive deer-like ruminant. Differs from other deer by not having antlers and facial glands. Has gall bladder, caudal gland and musk gland which no other deer possess. Canines are greatly developed especially in males. Stockily built animal with small head having hare-like long ears, arched back and bounding gait. Hind legs are 5 cm longer than the forelegs. Well-developed dew claws or lateral hooves. The general body colour is grey-brown to light-brownish yellow or dark yellow or dark brown with two white throat stripes. The colour of the ear tips range from dark brown to ivory-yellow or white. Rump and thigh are darker and the belly pale brown. Genital regions are white and the tail is naked except for a tuft of hair at the tip. Fur coat is thick and wavy giving it a deceptively heavy built appearance.

Musk deer is a Small Solitary Forest Ruminant (SSFR) that inhabits the upper temperate, subalpine and alpine habitats between 2,500 m and 'treeline'. It is a sedentary animal and is either nocturnal or crepuscular spending most of the day under dense undergrowth and come out of cover for feeding in open alpine grasslands during night but seldom stray far away from escape cover. Wild predators of musk deer include snow and common leopards, Himalayan yellow-throated marten, and possibly Red fox. It makes a non-vocal alarm call "Hiss" and stot on sensing danger. Man has persecuted this species for centuries for its musk. Being a SSFR, musk deer rely largely on communication by olfaction which is by means of scent marking behaviour and that includes defecating at defecation sites and by pasting with the secretions of the caudal gland by both sexes, and in case of males by the secretion of musk in urine. Musk deer has high-energy requirement per unit of body weight and it must select high quality food. It is called a "nibbler" rather than browser as it selectively feed on young leaves, buds, fruits, flowers of dicots, mosses and lichen.

KASHMIR STAG

Cervus elaphus hanglu Linnaeus 1758
Vernacular/other Names: hangul, honglu (Hindi/Kashmiri)
Taxonomy
Family: Cervidae
Subfamily: Cervinae

Conservation Status: At present occurs only in the Dachigam National Park and a few isolated pockets in Jammu & Kashmir State of India. In the past, it was distributed in mountains of Kashmir and in Chamba District of Himachal Pradesh. Subalpine forests, temperate broad leaved oak and conifer mixed forests, riverine forests, grasslands and scrub habitats

A subspecies of the Red deer of Europe that is distinctly different from the Red deer in body size, colour, and antler size and shape. Hangul is dark grey and dark brown rather than reddish and unlike Red deer, the antler bay tine is normally larger than brow and has much of cup. In Hangul, 10-point head is the normal one and any greater number of points is far more usual than in the case of Red deer. The coat colour ranges from light to dark brown, fading to dingy white on lips, chin, underparts and anal region. The white rump patch does not extend much above the tail and is divided by a broad median stripe extending down to the base of the tail and sometimes to its extremity. Coat colour fades during summer, but tones up with denser coat in winter, which in big males is very dark or rufous brown. Fawns are spotted. Old females may show white flecks. Antlers in adult males grow up to 1 m

Hangul is essential a forest animal found singly or in groups of about 10-15 individuals. It is diurnal and make distinct altitudinal migration from higher to lower elevations in winter. Predators are snow and common leopards. It is a selective feeder and feeds on a wide variety of plants that includes leaves/shoots of some trees and shrubs, grasses and herbs.

Another subspecies of the Red Deer that is reported in the Indian subcontinent is Shou or Sikkim stag (Cervus elaphus wallichi). It is reported to occur in Chumbi valley located close to the Indo-China border in Sikkim and there are unconfirmed reports of this subspecies being sighted occassionally in Pangolokha Wildlife Sanctuary

MAINLAND SEROW

Nemorhaedus sumatraensis Bechstein 1799

Vernacular/other Names: sarao, thar (nepali), ramu (Kashmiri), gya (Sikkim)

Taxonomy

Family: Bovidae

Sub Family: Caprinae

Tribe: Rupicaprini

Conservation Status

In the southern side of the Greater Himalaya (2,000-3,300m); and in the hills of N.E. India (>200 m). Occurs in the states of Himachal Pradesh, Uttaranchal, Sikkim, Arunachal Pradesh, Mizoram and Manipur. Temperate and subalpine forests in the Greater Himalaya and montane wet temperate and evergreen forests of N.E. India. Two subspecies of serow are reported to occur in India, the Black Serow (N.s.edwardsi) distributed in the western and eastern Himalaya and the Red Serow (N.s.rubidus) distributed in eastern Himalaya.

An elusive mountain goat, with a large head, donkey-like ears, thick neck and short legs. The coat is coarse and rather thin; it colour varies from grizzled black or blackish grey-roan to red.

In the darker males, the head, neck, and the mane that covers the nape and withers are grizzled black. The black passes into rusty red on the shoulders, flanks and lower thighs, and turns a dirty grey on the inside of the limbs and belly. The muzzle, throat and chest have varying amounts of white. Both sexes have horns which are black, conical and closely wrinkled for three quarters of their length

A solitary and shy mountain goat inhabiting steep, rugged, inaccessible densely forested habitats that includes thickly wooded-gorges with boulder-strewn slopes and shallow caves or overhangs. Although solitary in nature, serow is also found in pairs and in groups of three or four. Serow is a territorial species and appears to be nocturnal or crepuscular. Predators are the Common leopard and possibly Red fox and Himalayan yellow-throated Marten on the young. When disturbed, gives an alarm call that is a loud whistle. It feeds on a variety of food items including oak leaves, shrubs, grasses, montane bamboo, ferns, mosses and lichen.

TAKIN

Budorcas taxicolor Hodgson 1850

Vernacular/other Names: Not known

Taxonomy

Family: Bovidae

Sub Family: Caprinae

Arunachal Pradesh. Tropical forests (900-1,200m), dense bamboo and rhododendron forests (2,000-3,000m). Recently reported from Kyongnosla Alpine and Pangolkha Wildlife Sanctuaries in Sikkim.

A clumsy heavy animal, having a convex face, heavy mouth, and tremendously thick neck. The muzzle, except for a base spot at the extremity is covered with hair. Withers are slightly raised and its narrow back arches in the centre and slope downwards to the root of the tail. Body colour ranges from dark brown to golden yellow. Withers are conspicuously lighter in tone. Adult males are golden yellow merging into dark brown or black on the flanks and quarters. There is a dark dorsal stripe. Young males are reddish brown in front merging into black. Calves all black. Females are grayer than males with no yellow. The dark dorsal stripe is inconspicuous in females and young males. Horns of young Takin grow straight up from the head with an outward tendency. Later they grow outwards and then downwards. In final stages, the horns grow forward and bend downwards and outwards with the points growing up.

Inhabits the steepest and most thickly wooded habitats in the mountains. In summer, found in large herds of even up to 300 individuals, and break up into small parties in winter. Generally active during early morning and late afternoon, while resting in cover during mid day.

SPECIES OF THE TRANS-HIMALAYA AND OTHER ARID TRACTS

Rocky alpine steppe and the rugged tracts of the Central Asian mountains-from the Hindu Kush in Afganistan till the Sutluj gorge in Himachal Pradesh, India; and northwards along the Altai Shan and Tien Shan, into Mongolia and Russia. Within India in western Ladakh (Zanskar and Ladakh ranges) and Nubra in Jammu & Kashmir, and Lahul-Spiti in Himachal Pradesh. Covers ca. 25,000 km^2 within these two states.

Stocky built goat. Largest in the genus. The general colour of the pelage is a light tan, with the undersides lighter; the coat colouration however, varies widely across this ibex's range. In the Himalayan population during winter, mature males are much darker (dark chocolate brown; especially males older than 5 years), with patches of white on the neck and back. Dark dorsal stripe from lower neck till the tail pronounced in males.

The lighter ventral parts of females and young males are separated from the main body colour by a darker band which runs along the sides. Both sexes have a dark beard beneath the chin, although it is much less pronounced in females. Both sexes carry horns-in males they grow into massive arcs, which curl over the back and may even loop back on themselves, while in females they are small, almost parallel and slightly arched towards the rear. A male's horns have several large knobs on their frontal surface. Till the 7th-8th year two knobs are added each year; after which it is usually one every year.

The Siberian ibex live at high elevations, and often climb up to the vegetation line at 5,500m. In India, they usually occur between 2,500 and 5,500m. In many areas, they do not descend to lower elevations during winter. Uses primarily steep slopes, interspersed with 'escape terrain' in the form of rocky overhangs and cliffs, where they flee to escape predators such as snow leopard and the Tibetan wolf.

In winter, partial to south facing slopes and areas where snow deposition is relatively lower such as windblown ridgelines, crests and avalanche paths. When snow cover is heavy, Siberian ibex find food by pawing at the snow with their forelegs in order to reach the vegetation concealed below.

Most hierarchy is established before the onset of the rut based on body and horn size, but when contests do take place they may extend for over three hours of clashes, body and shoulder pushing and horn entwining.

During the clashes, a male would typically go upslope, rear on its hind limbs and clash down at the lowered horns of the opponent, often trying to dislodge it. Occur in small groups-mean around 11 animals per group; occasional congregations of 70-100 animals may occur in spring. Principally a diurnal animal, the Siberian ibex has alternating periods of resting and activity throughout the day. Usually forage in early forenoon and late afternoon, but have a distinctly unimodal foraging peak during the noontime in peak winters. Adult males primarily separate into 'all-male groups' during the summer.

Is an intermediate forager, feeding primarily on grasses, forbs, leaves and shoots of shrubs and also trees.

LADAKH URIAL

Ovis vignei vignei, Blyth 1841

Vernacular Names: Shapo (male), Shammo, Shanmar (female) (Ladakhi)

Taxonomy

Family: Bovidae

Sub family:Caprinae

Tribe:Caprini

Conservation Status: A subspecies endemic to the Ladakh region. Occurs is relatively less steep and rolling slopes usually below 4,200m-mainly along the Indus River and the Shyok River and their tributaries. Within India, occurs along the Indus almost continuously from Upshi in the east till Dha-Hanu; a side population also occurs in the Wanla-Lamayuru-Nindum area towards the south and in the Markha Valley. The Shyok population is mostly concentrated around Tirit-Sumur and some occur further upstream along the Nubra River and along the Diskit-Khalsar area. Distribution in India covers ca. 2,000km2. Small populations also reported from the Shigar valley, Gilgit and Skardu regions of PoK. Other related subspecies are distributed in the Kirthar, Safed Koh and Hindu Kush ranges in Pakistan and in parts of eastern Afghanistan and Tajakistan.

Smallest wild sheep species. Adult urial are grey-brown in winter coat, but are rufous grey or brown during the rest of the year. Adult rams sport a black or grizzled ruff, which grows from the side of the chin and meets below and extends down the throat. The ruff is more pronounced during winter and in males older than about 5 years. In adult rams, a dirty white and indistinct saddle patch is present. They have a dark spot under the foreleg that may extend on to a lateral stripe in some males. Females are similar, but lack the saddle patch and invariable have just a dark spot under their forelegs. Underparts and rump patch in all age-sex classes are white. Male horns curve up, back, downwards and in large males turn a bit inwards. Females have short, parallel horns and are not divergent as in bharal and argali. Grey patches on face from the eye till the chin is present in both sexes.

Ladakh urial are primarily animals of the lower slopes, usually below 4,200 m; rarely occur up to 4,800 m as in the proposed Nindum Wildlife Sanctuary.

Occur on gradual to moderate slopes, but have probably been pushed to use steeper and rockier gorges along the side valleys of Indus due to excessive anthropogenic pressures. Has long, lithe limbs that facilitate fast flight from potential danger. They are also known to retreat into cliffs or steep slopes to escape predators such as wolf and snow leopard. Usually occur in small groups of 5-12 animals, however, in certain areas and in seasons such as spring may occur in larger groups of ca. 80 animals. Sexual segregation is reported during much of the year barring the rutting period.

Is an intermediate forager, consuming grasses, forbs and shrubs in different seasons. During the rut, solitary males may travel vast distances in search of oesterous females. Contests between males are in the typical form of sheep, where two opposing rams clash into each other from some distance producing a loud clack. They may then involve in shoulder pushing and horn entwining.

TIBETAN ARGALI

Ovis ammon hodgsoni, Blyth 1840
Vernacular Names: Nayan, nyan (Ladakh, Tibetan, Spiti)
Taxonomy
Family: Bovidae
Sub family: Caprinae
Tribe: Caprini

Conservation Status: Open, rolling high altitude slopes and plateaus in Central Asia. Different subspecies are spread from the Pamirs in the west to the eastern Tibetan plateau; and from the areas just north of the Greater Himalaya, to Mongolia, in the north. In India less than 300 might be surviving in its range in Ladakh with smaller portions of the range also covered in Sikkim, totaling to ca. 12,000 km^2.

Largest wild sheep in the world. Long limbs, with a pronounced shoulder and a large chest. The general colouration of argali is variable, ranging from a light buff to darker grey-brown in winter. The underparts are whitish, usually separated from the darker parts by a blackish stripe. In addition, males have a whitish neck ruff, and a dorsal crest, both of which are more prominent in the winter coat. Argali has a whitish, large rump patch. Adult males carry two enormous, ribbed, horns, which turn upwards, back, down, and then forward and outwards and can reach 145-190 cm when measured along the spiral. The basal circumference of the horns may measure up to 40cm. Females also bear horns, although these are much smaller, rarely exceeding 30 cm in length. These are however thicker and more divergent than other sympatric caprine female horns.

Usually occurs at the highest elevations in the Tibetan Plateau (4,500-5,500m), using the open mesic alpine meadows. While local migrations may occur in autumn and spring, argali may not descend much in winters, but may chose areas with less snow (they primarily occur is regions where winter precipitation is naturally not very high). With relatively long legs, argalis are fast runners and flee away from their primarily cursorial predators, the Tibetan wolf.

However, refuge is occasionally taken on steep mountain slopes. The primary vocalizations are an alarm whistle and a warning hiss made by blowing air through the nostrils. When competing, males rear up on their hind legs and, leaning forward, race towards their opponent, crashing horns in the process.

Occur usually in small groups 2-15 animals, but occasional aggregations of about 40 animals also reported. Herds are sexually segregated for most of the year, except the rutting season. Is an intermediate forager feeding on herbs, sedges, grasses and shrubs. Some recent studies suggest that they rely to a greater extent on forbs and shrubs.

TIBETAN WILD ASS OR KIANG

Equus kiang
Vernacular Names: Kiang (Ladakhi, Tibetan)
Taxonomy
Family: Equidae
Sub family:-
Tribe:-

Conservation Status: Occurs all along the rolling mountains and valleys of the Tibetan Plateau between 4,000 and 5,200 m. Distributed from Ladakh in the west to the Quinghai province of China in the east; and from the plateau regions on the northern fringe of the Greater Himalaya, till the Xinjiang province of China. In India, kiang are reported from the Chang Thang plateau of Ladakh and the northern cold desert areas of Sikkim.

The largest of the wild asses, the kiang is a robust animal. The head is large, the muzzle blunt and the nose convex. The mane is upright and relatively short. The coat is a rich chestnut colour or dark brown in winter, turning to reddish brown in summer. The legs, the ventral part of the body, including the neck, inside of the ears and the muzzle are white. A dark dorsal stripe extends from the neck till the base of the tail. The tips of the ears and a narrow band along the hooves are also black. Little sexual dimorphism exists in the species, but the males are slightly larger than females.

Occurs usually from 4,000 m till the limit of vegetation at ca. 5,200 m, but within this range it is largely confined to flat plains, plateaus and rolling smooth slopes. Kiang use alpine meadows and alpine steppes dominated by grasses and sedges, but also occur on drier desert steppes.

Kiang may show minor seasonal migrations, most of it being altitudinal-migrating to higher regions in summer and to lower ones in winter. Within the Indian region, kiang are often seen in the moister wide river valleys and lake basins of the Chang Thang during the winter; an area also used by the nomadic domestic sheep-goat herders of the region. Kiang is large and swift-footed animal and predation on adults by wild carnivores is rare.

The natural predator of the region is the Tibetan wolf; and man also takes a toll-hunting of kiang is still known to be prevalent in Tibet, but not in the Indian parts of its range. On approaching, kiang typically flee with its head up and wheel around after running 80-100 m to look back at the potential danger. Often, they appear inquisitive and begin approaching the intruder.

Kiang normally occurs in small groups of 8-10 animals. Solitary animals, mostly stallions are also common. Occasionally, they may form large aggregations of 200-300 animals. Such aggregations are most common during autumn and winter when animals congregate on meadows with better forage quantity. Males tend to maintain small temporary harems during the rutting period and may rarely also show site fidelity suggesting some territoriality in the species.

Kiang are 'bulk foragers' and require larger quantities of forage, but not necessarily of good quality. Thus, it survives on relatively poor quality diet of sedges and grasses during the entire year.

TIBETAN ANTELOPE

Pantholops hodgsonii, Abel 1826

Vernacular Names: Tstosh (Ladakhi, Tibetan), chiru (parts of Tibet)

Taxonomy

Family: Bovidae

Sub family: Panthalopinae

Tribe: Saigini

Conservation Status: The Tibetan steppe at elevations of 4,000-5,500 metres in open, plains and rolling mountains. Occurs in the Chang Thang plains in northern Tibet, Xinjiang and Qinghai. Smaller populations occur in central Tibet. Small groups of female and young from the northern Chang Thang population from Tibet are known to migrate into Indian territories in the Aksai Chin (Daulat Beg Uldi-Lingti-Tsiang plains) and males in the Changchenmo valley during June till autumn. The combined distribution range within India will be less than $300km^2$.

The dense coat is very soft and woolly, and is very good insulation against the cold Tibetan weather. The coat of adult males is reddish fawn with light grey and brown tones grading to creamy white undersides.

The fronts of the slender legs and the face (including the forehead, bridge of the nose, and upper cheeks) are dark brown. By October the males acquire the more striking winter pelage. The face and the front of the legs are now black, and the marking from the foreleg goes over the shoulder and from the hindlegs extends around the margins of the rump patch.

The rest of the body is light gray and tan. Females are smaller, and the coat is fawn-almost pinkish, often with rust brown in the nape that blends into the lighter underside. Males alone carry the slightly S-shaped horns which grow 50-70 cm in length. Black in colour and ridged on their lower half, they rise nearly vertically from the head. Another distinctive feature is the presence of walnut sized airsacs on each side of the nostril, that gives the muzzle a blunt and enlarged look. The fawn-coloured underwool or shahtoosh produced by this species is the finest wool known to man.

TIBETAN GAZELLE

Procapra picticaudata, Hodgson, 1846
Vernacular Names: Gowa (Ladakhi, Tibetan)
Taxonomy
Family: Bovidae
Sub family: Antilopinae
Tribe: Antilopini
Conservation Status
IUCN: Low risk
CAMP: Critical
IWPA: Schedule I
CITES: Not listed
US ESA: Not listed

Open, rolling plateaus, slopes and valleys, with mesic alpine meadow/steppe vegetation communities, usually above an elevation of 4,200 m. Global range extends in much of the Tibetan Plateau and the Xinjiang, Quinghai and Gansu provinces of China. In India, small populations occur in 3, possibly 4 places in eastern Ladakh (Hanle, Chushul, Chumur and Changchenmo) and northern parts of Sikkim (near Gyaum Chhona and Changri meadows). The overall occurrence may be in less than 500 km^2 in the country.

The gazelle is a small animal with slender legs weighing between 13 and 15kg. Its coat varies from sandy brown to grayish brown, and is greyer in summer.

The underparts and insides of the legs are white. A prominent feature is the large white heart shaped rump patch, which covers much of the rump. A rust-brown band surrounds the rump patch. Only males have horns. These slender horns go up, curve back and then up again. They are ridged till the distal quarter and are slightly divergent at the top and may again bend inwards.

Species of the open plateaus and mountains occurring between elevations of 4,200 and 5,500m. Primarily occur in the mesic alpine meadows and steppe vegetation. Rarely also occurs in desert steppes. Is a lithe animal that relies on quick detection and flight from predators like the Tibetan wolf and, and rarely wild dog and snow leopard.

Little altitudinal migration is reported, but generally occur in their lowest range in the spring. Winter is a hard time for this small ungulate, especially in years with heavy snowfall when heavy mortality may occur.

The gazelle usually occurs in small groups of 4-5 animals, and rarely larger concentrations may occur. In India the total population is likely to be less than 50 in three small populations and seeing large groups any where is extremely difficult. During rutting, males may hold small territories where they maintain small harems. These sites are often marked by 'laterine sites'.

The gazelle tends to be a 'concentrate forager' feeding on more nutritious forbs and lush sedges during all periods, except winter, when it switches to the only forage which is coarse for its survival.

YAK

Bos grunniens, Linnaeus 1766
Vernacular Names: Dong (Ladakhi, Tibetan)
Taxonomy
Family: Bovidae
Sub family:-
Tribe: Bovini
Conservation Status
IUCN: Vulnerable
CAMP: Critical
IWPA: Schedule I
CITES: Not listed
US ESA: Endangered

Montane regions of Tibet at an altitude of 4,000-5,400m on relatively moist grass and sedge dominated slopes and valleys. Distribution is somewhat restricted compared with other Tibetan ungulates to the northern parts of Tibetan Chang Thang, extending into the Xinjiang province in the north and the western extremity of the Quinghai province. Good populations are reported from the Aru Basin in the Changthang Nature Preserve. In India, the distribution is restricted to small portions in the Changchenmo valley in Ladakh, just north of the Pongong Tso where some 40 animals may exist.

The dark black-brown coat is dense, wooly, and extremely shaggy, although the colour is variable in domestic yaks. 'Golden' wild yaks are also known, but are rare. The shoulders are high and humped, with a broad, drooping head.

The short legs have broad hooves and large dewclaws as an adaptation to mountainous environments. Long mantles of hair drape down from the neck and flanks, often almost completely covering the legs. The tails are large and bushy. The horns are found in both sexes, though those of the females are considerably smaller and shorter. The curved horns grow out from the sides of the head and curve upwards. The wild yak horns are on an average larger than the domestic ones and have a distinctive white splashes on the muzzle. Occur at high elevations (4,000 to the limit of vegetation at ca. 5,400m) on mesic to moist meadows and rarely on more arid areas. Prefers rolling slopes and flat plains or plateaus. Coarse/bulk forager requiring lot of forage every day, it survives on grasses, sedges, and occasionally lichens and tubers. Due to the scarcity and unpredictability of vegetation in their habitat, wild yaks must travel great distances in order to obtain sufficient nourishment. In July, there is a general downward shift from the high plateaus to the lower plains, where swamp vegetation is at their peak growth.

As the temperature rises in August, the herds head back up to the plateaus, even retiring to snowy regions to beat the heat. Although sensitive to warm temperatures, wild yaks can easily tolerate temperatures of-40°C. Generally wary, if a herd is disturbed they will flee for a long distance, galloping with their tails held erect.

The only natural predator is the Tibetan wolf, but may have limited success of killing an adult yak. When cornered, or when young are present in the group, yaks may actively defend them against any potential danger.

In spite of their bulky, awkward appearance, yaks are excellent, sure-footed climbers. One of the few vocalizations is a loud grunt, made during the breeding season by wild yaks. Domestic yaks, however, "grunt" throughout the year-hence the specific name 'grunniens'.

Solitary animals and herds of 10-50 are most common; occasionally 200 individuals are also seen. Aggregations of over 80 animals are commonly seen in some areas such as the Aru basin in Tibet.

FOREST PEST SPECIES INFLUENCED BY CLIMATE CHANGE

Some examples of forest insect pests, diseases and other pests which have been impacted or are predicted to be impacted by climate change are presented below. Information on non-forest pests is also provided to enable a better understanding of the potential impacts of climate change on forest health.

HEMIPTERA

Aphids (Aphididae)

With short generation times and low developmental threshold temperatures, aphids are a group of insects that can be expected to be strongly influenced by environmental and climatic changes. In general, it has been predicted that aphids will appear at least eight days earlier in the spring within 50 years, though the rate of advance will vary depending on location and species. This could potentially result in greater damage to host plants depending on the phenology of host plants and natural enemies.

Zhou *et al.* (1995), for example, investigated the timing of migration in Great Britain for five aphid species (*Brachycaudus helichrysi, Elatobium abietinum,*

Metopolophium dirhodum, Myzus persicae, Sitobion avenae) over a period of almost 30 years and concluded that temperature, especially winter temperature, is the dominant factor affecting aphid phenology for all species. They found that a one degree Celsius increase in average winter temperature advanced the migration phenology by 4-19 days depending on species.

Elatobium abietinum (Walker) (Aphididae)

The green spruce aphid (*Elatobium abietinum*) is also believed likely to benefit from the increase in winter survival, leading to more intense and frequent defoliation of host spruce trees (*Picea* spp.). This aphid is native to Europe but has also been reported in both North and South America.

Infestations in Great Britain have resulted in large losses of spruce foliage and height both during the active infestation and in subsequent years. Westgarth-Smith *et al.* (2007) showed that warm weather associated with a positive North Atlantic Oscillation (NAO) index caused spring migration of *E. abietinum* to start earlier, last longer and contain more aphids. Positive NAO values correspond to warmer atmospheric conditions over Great Britain. Since global warming is believed to increase NAO variability, shifting the system to more positive values, this will most likely lead to further increases in aphid activity and more damage to spruce trees and forests in the area.

Coleoptera

Agrilus Pannonicus

A number of Buprestid beetles of the genus *Agrilus* have been linked to oak decline. Incidences of these species have increased worldwide (both in their countries of origin and by international movement) and their impacts are being linked to host tree stress potentially caused by climate change. For example, *Agrilus pannonicus* (=*A. biguttatus* (Fabricius)) has recently been associated with a European oak decline throughout its natural range and has increased in incidence in several countries including France, Germany, Hungary, Poland and the Netherlands, and the UK where it is believed to be contributing to oak decline. Infestations can result in extensive tree mortality which, combined with other factors involved in the decline, can drastically alter the species composition of oak forests.

Dendroctonus Frontalis Zimmermann

Dendroctonus frontalis is considered to be one of the most damaging species of bark beetles in Central America and southern areas of North America. It is a major pest of pines and has a wide distribution occurring from Pennsylvania in the United States south to Mexico and Central America. Populations can build rapidly to outbreak proportions and large numbers of trees are killed.

Initial attacks are generally on weakened trees however *D. frontalis* is capable of killing otherwise healthy trees. This beetle kills trees by a combination of two factors: girdling during construction of egg galleries; and the introduction of blue stain fungi of the genus *Ophiostoma.* Because of their short generation time, high dispersal abilities and broad distribution of suitable host trees, the southern pine beetle has the potential to respond quickly and dramatically to any changes in climate.

In October 1998, Hurricane Mitch hit Central America, causing floods and mudslides that ravaged local communities, forests and infrastructure. In the years that followed an unprecedented regionwide outbreak of pine bark beetles, mainly *D. frontalis* in association with other *Dendroctonus* and *Ips* species, destroyed over 100 000 ha of pine forest. As most of the standing dead and felled trees were left on site, fuel loads were drastically increased thereby resulting in extensive wildfires. With climate change expected to increase the frequency and severity of extreme events such as hurricanes, the potential for future devastating impacts on forests from both the initial disturbance and its cascading effects (*i.e.* other disturbances such as pest outbreaks and fire) is quite high.

Warmer temperatures attributed to climate change have also influenced the southern pine beetle resulting in range expansions in the United States. Laboratory measurements and published records of mortality in wild populations indicate that a temperature of -16 °C or less result in almost 100 percent mortality of the pest, thereby limiting its distribution in its current northern range. It was predicted that an increase in temperature of 3 °C would enable outbreaks to occur approximately 178 km farther north than in historical times (Ungerer, Ayres and Lombardero, 1999). Recent outbreaks of the southern pine beetle in northern and high-altitude ecosystems, where they were previously rare or absent, have been attributed to a warming trend of 3.3 °C in minimum winter air temperatures in the southeastern US from 1960-2004. This northern expansion is about as predicted by Ungerer, Ayres and Lombardero (1999).

The southern pine beetle has also possibly adapted ways to increase survival in cooler climes. Tran *et al.* (2007) showed, through field and laboratory studies of a northern population, that prepupae were more cold tolerant (by more than 3 °C) than pupae, adults and feeding larvae, and that the winter life stage structure was strongly biased toward this most cold-tolerant life stage. This tendency to overwinter in a cold tolerant life stage could be a coincidence however rather than a true adaptation.

Dendroctonus Ponderosae

The mountain pine beetle (*Dendroctonus ponderosae*) is the most destructive pest of mature pines in North America, particularly lodgepole pine (*Pinus contorta*). In the western United States, outbreaks have been increasing in area after several years of drought (Tkacz, Moody and Villa Castillo, 2007). A major epidemic of this pest has also been ongoing in western Canada (British Columbia (BC), and more recently, Alberta) for several years and even with large-scale efforts to mitigate the impacts of the pest, millions of trees have been killed. A record of over 10 million hectares of pines were recorded as infested during 2007 aerial overview surveys in BC, with 860 973 ha of this located in provincial parks and protected areas. It has been predicted that if the beetle continues to spread at its current rate as much as 80 per cent of mature pine in BC will be dead by 2013. The large numbers of dead and dying trees have also increased the risk of wildfires.

The problem has been exacerbated by successive years of mild winters, resulting in decreases of mortality of overwintering stages and generation time. Their life cycle is generally completed in one year; warmer temperatures can result in two generations per year while cooler ones may results in one generation every two years (Amman, McGregor and Dolph, 1990). Drought conditions associated with warmer temperatures have also weakened the trees and increased their susceptibility to the beetles. Warmer temperatures have thus opened up previously climatically unsuitable mature pine stands to the pest.

Dendroctonus Rufipennis

Dendroctonus rufipennis is a North American pest of spruce, particularly white spruce (*Picea glauca*) and black spruce (*P. mariana*) in the north, Engelmann spruce (*P. engelmannii*) and sitka spruce (*P. sitchensis*) in the west, and red spruce (*P. rubens*) in the east. It tends to attack weakened or windthrown trees and outbreaks are mostly linked to predisposing factors. As a result it can be expected that the impacts of climate change on trees and forests could enhance spruce beetle outbreaks.

In fact, Hebertson and Jenkins (2008) investigated the impact of climate on spruce beetle outbreaks in Utah and Colorado, USA between 1905 and 1996 and found that historic outbreak years in the intermountain region were related to generally warm fall and winter temperatures and drought conditions. Similarly outbreaks in both Canada (Yukon Territory) and the US (Alaska) appear to be related to extremely high summer temperatures which influenced spruce beetle population size through a combination of increased overwinter survival, a halving of the maturation time from two years to one year, and regional drought-induced stress of mature host trees.

Hymenoptera

Cephalcia arvensis Panzer

The spruce webspinning sawfly is monophagous on spruce (*Picea*) and endemic to the spruce range in Eurasia, where outbreaks have been seldom recorded. From 1985-1992 however there was a sudden outbreak of the sawfly in the Southern Alps during which populations developed an annual life cycle and grew exponentially, causing repeated defoliations resulting in extensive tree death.

Cephalcia species generally show low fecundity and have an extended diapause of a few years that is stimulated by low temperatures at pupation time. The outbreak corresponded to a period of high temperatures and low precipitation and severe water stress for the host trees.

As a result, the insect was able to adapt to the new climate resulting in lower mortality, faster development and higher feeding rates of the sawfly. In addition, the sudden increase in population density was not quickly followed by that of natural enemies, thereby allowing for unlimited population growth.

Neodiprion Sertifer

The European pine sawfly *Neodiprion sertifer* is an important pest species on pines in Europe, northern Asia, Japan and North America where it was introduced. It is one of the most serious defoliators of Scots pine (*Pinus sylvestris*) forests in northern Europe. Virtanen *et al.* (1996) suggested that outbreaks of the sawfly on Scots pine in eastern and northern Finland are prevented by low winter temperatures which kill eggs, and predicted that outbreaks would become more common with winter warming. A high variation in freezing avoidance of eggs was also noted which would allow *N. sertifer* to adapt to predicted climate change and spread its distribution northwards.

Ascomycota

Mycosphaerella pini Rostrup

Mycosphaerella pini is a fungus that infects and kills the needles of *Pinus* spp. resulting in significant defoliation, stunted growth and eventually death of host trees although susceptibility among pine species does vary. Native to temperate forests of the northern hemisphere, it is a major pest of pine plantations in the Southern Hemisphere, where both the host and the pathogen have been introduced. The widely planted *P. radiata* is particularly susceptible and many forests planted with this species in the Southern Hemisphere, particularly in East Africa, New Zealand and Chile, have been devastated by this needle blight. This pathogen has forced managers in some areas to abandon the planting of *P. radiata* and depend more on other tree species.

In its native range *M. pini* normally causes little damage, but since the late 1990s it has been causing extensive defoliation and mortality in young plantations of lodgepole pine (*Pinus contorta* var. *latifolia*) in northwestern British Columbia, Canada (Woods, Coates and Hamann, 2005). Mortality of mature lodgepole pines has been observed in mixed-species stands, where scattered pine represents only a small proportion of stand composition; this represents a globally unprecedented occurrence for *M. pini* (Woods, Coates and Hamann, 2005). The current epidemic coincides with a prolonged period of increased frequency of warm rain events throughout the mid-to-late 1990s allowing for the rapid spread and increased rates of infection. Unlike many other pests, changes in precipitation patterns may be more important than changes in temperature for predicting the spread and impact of *M. pini*.

The Climates of the Past and Present

The planet on which we live is an oblate spheroid, its equatorial region being warmer than the two polar areas. This temperature gradient, in conjunction with the earth's rotation, sets up a reasonably standard pattern of atmospheric circulation. Since the surface of the earth is neither all land nor all water, but divided into land masses surrounded by even larger areas of ocean, the physical properties of water tend to decrease seasonal contrasts near continental margins relative to those in interior regions. Any enlargement of continental areas would tend to increase the seasonal contrasts, induce steeper barometric gradients and thus heighten the general storminess of their interiors.

Conversely, any decrease in the continental areas, as by a lowering of their general masses with an encroachment of oceanic waters, would tend to bring about an amelioration of climate and a general stabilization of seasonal variations. As the geological record clearly indicates, there have been times when the parts of the continental masses exposed above oceanic waters increased for a while and then shrank, with long periods of relative quiescence between, often accompanied by extensive invasions of the sea.

The periods of increase of the exposed portions of the continents were, on the whole, correlated with periods of mountain building and the development of plateaus. The geological record also indicates that these periods of continental uplift were usually accompanied by intermittent glaciation, at least in the polar regions. Volcanic activity of greater or lesser intensity, especially near the continental margins, also was characteristic of such times.

The fact that we happen to be living in one of these cycles of disturbance focuses our attention on the climatic (and consequent vegetational) features of such periods rather than on the more quiescent and far longer periods that have intervened between them. As Russell has pointed out, man during his entire history has not known what it is to live in a "normal" climate. One may also add that, during, his existence, man has not witnessed what might be termed a "normal" phytogeographic pattern.

If the surface of the earth were uniform there would be three parallel zones of higher-than-average precipitation, one in the equatorial region and the other two in the general regions of the 60° parallels of latitude. There also would be four zones of lower-than-average precipitation, two in the polar regions and two centering about 30° N and S latitude. Because of the rotation of the earth and the interaction of factors resulting from the positions of continental masses and oceans — and their unequal distribution — this theoretical pattern is greatly altered. Furthermore, as Rossby has noted, in the northern hemisphere the much greater percentage of land tends to disrupt the expected pattern more than it is disrupted in the southern hemisphere with its much larger areas of ocean.

There is a close correlation between effective precipitation and vegetation. Hence one may draw certain inferences regarding precipitation from a generalized map of the world's vegetation types. In spite of the heat, the precipitation near the equator is such that few desert areas are found there. Apart from polar wastelands, subhumid and essentially treeless regions tend to develop at the western continental margins at about latitudes 20°-30° and to trend inland in a northeastward direction in the northern hemisphere and southeastward in the southern hemisphere. Although local conditions such as mountain ranges modify these relationships, this, on the whole, is the general pattern.

The belts of heavy precipitation to be expected on a uniform-surfaced world at about 60° would also be turned northeastward on the continents of the northern hemisphere and southeastward in the southern hemisphere. In the skewing of these belts they would be depressed about 100 at the western continental margins. Their poleward trends would soon carry them into the polar subhumid areas; thus they would disappear as belts and become more or less regional coastal areas centering at about latitude 50°. Evidences of these are to be found in the northern hemisphere in the areas of heavy precipitation in North America in the Pacific Northwest and also in northwestern Europe. In the southern hemisphere only South America extends into sufficiently southerly latitudes for the southern superhumid area to be fully evident, as in southern Chile; but here no extended area is found, because of the southern Andes and the small total amount of land area available.

Reference to the vegetation map will indicate that the tropics, which we sometimes think of as characterized by heavy precipitation and an extensive development of rain forest, are by no means uniformly so. The less humid areas just north and south of the equatorial regions restrict the band of equatorial rain forest and other factors shift it and break it into segments.

Those areas, loosely referred to as "the tropics," actually are not characterized by a heavy effective precipitation and dense rain forest; the latter is a forest type of relatively limited extent. The montane rain forests sometimes developing on the windward slopes of mountains in frost-free regions are superficially similar to lowland rain forests, but the two would be distinguished in any detailed analysis of forest types. On the basis of theoretical climatology, those longer periods in the history of the world when its surface was relatively quiescent, with the exposed portions of the continents smaller than at present and also with somewhat subdued relief, were times of "normal" climate. Vegetational zonation poleward from the equator was evident, but without the sharp gradients of today. The air may have been slightly more humid than it now is over large areas, but the total rainfall probably was somewhat less.

The latter fact would tend to decrease the extent of the expected "standard" areas of superhumid climate, which were probably little better developed than they now are. Contrarily, with a more uniform air circulation, the subhumid zones entering the western margins of the continents, while in evidence, probably did not give rise to desert conditions except in the interiors of the largest of the land masses. Again on the basis of climatological considerations, it can be postulated that, during periods of world-wide continental depression and quiescence, the frost-free area of the world extended much farther poleward than at present. Theoretically the frost-free area would have reached to the neighborhood of latitudes 60° N and S; actually, judging from plant remains, it may have extended somewhat farther poleward on occasion. However, where one places the frost-free limits in relation to the 60° parallels depends to a certain extent on whether one does or does not believe in the theory of continental displacement.

Lastly, on purely theoretical grounds, the polar regions of such periods would be characterized by a cool but not cold climate. Snow probably fell in winter, but along the margin of any polar sea which existed the climate is not likely to have been much more severe than it now is in southwestern British Columbia, northeastern Nova Scotia, northern Scotland, or southern New Zealand. The abundant plant fossils of both the Arctic and Antarctic regions offer ample testimony that these theoretical speculations are not far wrong; if theory errs, the plants indicate that it may be on the side of conservatism.

4

Forest Policy Implementation in Modern World

THE NATIONAL FOREST POLICY

The forest policy of the United States may be said to have had its origin in 1799 in the enactment of a law which authorized the purchase of timber suitable for the use of the navy, or of land upon which such timber was growing. It is true that laws were in force under the early governments of Massachusetts, New Jersey and other colonies, providing for the care and protection of forest interests in various ways, but these laws were distinctly survivals of tendencies acquired in Europe and for the most part of little use.

It was not until the apparent approach of a dangerous shortage in certain timber supplies that the first real step in forest policy was taken by the United States. Successive laws passed from 1817 to 1831 strove to give larger effect to the original enactment, but without permanent influence towards the preservation of the live oak, which was the object in view. A long period of inaction followed these early measures. *The unshaded areas are treeless, except along the Streams* 1831 the solicitor of the treasury assumed a partial responsibility for the care and protection of the public timber lands and in 18 55 this duty was transferred to the commissioner of the general land office in the Department of the Interior.

The effect of these changes upon forest protection was unimportant. When, however, at the close of the Civil War railway building in the United States took on an unparalleled activity, the destruction of forests by fire and the axe increased in a corresponding ratio and public sentiment began to take alarm.

Action by several of the states slightly preceded that of the Federal government, but in 1876 Congress, acting under the inspiration of a memorial from the American Association for the Advancement of Science, authorized the appointment of an officer under the commissioner of agriculture, to collect and distribute information upon forest matters. As the railways advanced into the treeless interior, public interest in tree-planting became keen. In 1873

Congress passed and later amended and repealed the timber culture acts, which granted homesteads on the treeless public lands to settlers who planted one-fourth of their entries with trees. Though these measures were not successful in themselves they directed attention towards forestry. The act which repealed them in 1891 contained a clause which lies at the foundation of the present forest policy of the United States.

By it the president was authorized to set aside " any part of the public lands wholly or in part covered with timber or undergrowth, whether of commercial value or not, as public reservations and the President shall, by public proclamation, declare the establishment of such reservations and the limits thereof." Some eighteen million :acres had been proclaimed as reservations at the time when, in 1896, the National Academy of Sciences was asked by the secretary of the interior to make an investigation and report upon " the inauguration of a rational forest policy for the forest lands of the United States." Upon the recommendation of a commission named by the Academy, President Cleveland established more than twenty-one million acres of new reserves on the 22nd of February 1897. His action was widely misunderstood and attacked, but it awakened a public interest in forest questions without which the rapid progress of forestry in the United States since that time could never have been made.

Within a few months after the proclamation of the Cleveland reserves the present national forest policy took definite shape. Under this policy the national government holds and manages, in the common interest of all users of the forests or its products, such portions of the public lands as have been set aside by presidential proclamation in accordance with the act of 1891.

These lands are held against private acquisition under the Homestead Act (except as to agricultural lands as hereafter mentioned), the Timber and Stone Act and other laws under which the United States disposes of its unappropriated public domain, but not against private acquisition under the Mineral Land Laws. They are selected from lands believed to be more valuable for forest purposes than for agriculture and are managed with the purpose of securing from them the best and largest possible returns, present and future, whether in the form of water for irrigation or power, of timber, of forage for stock, or of any other beneficial product. The aggregate area of the reserves, or national forests, has been steadily increased until they now include nearly all the timber lands left of the public domain.

The general lines of this policy were in part laid down by the commission already mentioned, in its report submitted to the secretary of the interior, May 1, 1897 and by the act of June 4, 1897, which was largely shaped by the work of the commission. Until this act was passed the national forests had been in theory closed against any form of use; nor had the possibility of securing forest preservation by wise use received much thought from those who had favoured their creation. Such a state of affairs could not continue. Before long public opinion would have forced the opening to use of the resources thus :arbitrarily locked up and in the absence of any administrative system providing for conservative use, the national forests would inevitably have been abolished and the whole policy of government forest holdings would have ceased.

The act of June 4, 1897 was therefore of the first importance. This act conferred upon the secretary of the interior general powers for the proper management of the national forests through the general land office of his department. It provided for the designation and sale of dead, mature and large timber; authorized the secretary to permit free use of timber in small quantities by settlers, miners and residents; empowered him to "make such rules and regulations and establish such service as will insure the objects of such reservations, namely, to regulate their occupancy and use and to preserve the forests thereon from destruction "; and made violation of the act or of such rules and regulations a misdemeanour. The statute limited the power to establish forest reservations to the purpose of improving and protecting the forest, securing favourable conditions of water flows and furnishing a continuous supply of timber for the use and necessities of citizens of the United States.

Lands found, upon due examination, to be more valuable for other purposes than for forest uses might be eliminated from any reservation and all mineral lands within the reservations were left open to private appropriation under the mineral laws. The rights of settlers and claimants were safeguarded and civil and criminal jurisdiction, except so far as the punishment of offences against the United States in the reservations was concerned, was reserved to the States.

While the administration of the national forests was entrusted to the general land office, the same act assigned the surveying and mapping of them to the United States Geological Survey, which has published descriptions and maps of some of the more important.

No attempt was made in the general land office to develop a technical forest service. There were, indeed, at the time of passage of the act, less than ten trained foresters in the United States, no means of training more and very little conception of what forestry actually meant. The purpose of the administration was therefore mainly protection against trespass and fire, particularly the latter.

Regulations were made giving effect to the provisions of the act of June 4, set forth above, but in the absence of technical knowledge as to what might safely be done, the tendency was rather to restrict than to extend the use of the forest. Meanwhile, however, there was rapidly developing in another branch of the government service an organization qualified for actual forest management.

One year after the passage of the act of June 4, 1897, the division of forestry in the Department of Agriculture ceased to be merely a bureau of information and became an active agency for introducing the actual practice of forestry among private owners and for conducting the investigations upon which a sound American forest practice could be based.

The work awakened great interest among forest owners and exerted a powerful educational influence upon the country at large.

The division extended its work and became the Bureau of Forestry. It drew into its employment for a time nearly all the men who were preparing themselves in increasing numbers (at first abroad, then in the newly-founded schools in the United States) for the profession of forestry and was soon recognized as qualified to speak authoritatively on technical questions connected with the administration of the national forests.

This led to a request from the secretary of the interior for the advice of the bureau on such questions. Working plans were accordingly undertaken for a number of the forests. The general land office, however, was not ready to attempt active forest management. Though some timber was sold and the grazing of stock regulated to some extent, the main object of the land office administration continued to be protection against fire. Many of the regulations which it made could not be enforced.

The disadvantages of dispersal of the Federal government forest work among three separate agencies grew more and more apparent, until, on the 1st of February 1905, control of the 63,000,000 acres of forest reserves which up to that time had been set aside was transferred from the general land office to the Bureau of Forestry. In recognition of its new duties the designation of the bureau became the Forest Service.

Other provisions of the act which affected the transfer were that forest supervisors and rangers should be selected, so far as possible, from qualified citizens of the state or territory in which each forest was situated and that all money received from the sale of any products or the use of any land or resources of the national forests should be covered into the treasury and constitute a special fund for their protection, administration, improvement and extension.

Five days later a statute gave forest officers the power to arrest trespassers; and on the 3rd of March the lieu land selection law was repealed. This law had opened the way for grave abuses through the exchange of worthless land by private owners within the forests for an equal area of valuable timber lands outside.

The law has been modified since by the change of the old name " Forest Reserves " to " National Forests." The act of June iz, 1906, opened to homestead entry lands within national forests found by examination to be chiefly valuable for agriculture. The administration and improvement of the national forests are now provided for directly by congressional appropriation. The power to create national forests conferred on the president by the act of March 1891 has been repealed for the states of Washington, Oregon, Idaho, Montana, Wyoming and Colorado, but for no others.

The Forest Service began in earnest the development of all the resources of the national forests. Mature timber was sold wherever there was a demand for it and the permanent welfare of the forests and protection of the streams permitted, but always so as to prevent waste, guard against fire, protect young. growth and ensure reproduction. Regulations were adopted which allowed small sales to be made without formality or delay. secured for the government the full value of timber sold and eliminated unnecessary routine.

Care was taken to safeguard the interests of the government and provide for the maintenance of good technical standards. The conduct of local business was entrusted to local officers. Large transactions with general policies were controlled from Washington, but with careful 2 = *District Boundaries and Numbers* provision for first-hand knowledge and close touch with the work in the field.

Business efficiency and the convenience of the public were carefully studied. In short, an organization was created capable of handling safely, speedily and satisfactorily the complex business of making useful a forest property of vast extent, scattered through sixteen different states of an aggregate area of over 1,50o,000 sq. m. and with a population of 9,000,000.

The growth since the 1st of July 1897 of the area of the national forests, of the expenditures of the government for forestry and of the receipts from the national forests, is shown by the statement which follows. Though the act of June 4, 1897, became effective immediately upon its passage, the fiscal year 1899 was the first of actual administration, because the first for which Congress made the appropriation necessary to carry out the law.

Forest is ripe for the axe, the demand is strong and control by trained men makes it safe to cut more freely. The increase is marked both in small and in large sales, but a score of sales for less than $5000 are made against one for more. The total cut is still far below the annual increment of the forests. As the demand grows restrictions must increase in order to husband the present supply until the next crop matures.

- The stumpage price would seem on the face of the figures to have risen from about one dollar to more than three dollars per thousand board-feet. The receipts, however, for any one year are not exclusively for the timber cut in that year, since payments are made in advance. In the year 1907 the average price obtained was something less than $2.50 per thousand. It is therefore true that stumpage prices have risen greatly, although conditions new to the American lumbermen are im *Area of National Forests, Annual Expenditures of the Federal Government for Forestry and National Forest Administration and Receipts from National Forests, 1898-1909* Until 1906, the sole source of receipts was the sale of timber. In the fiscal year 1907, however, timber sales furnished less than half the receipts. The following statement concerning the timber sales of the fiscal years 1904-1907 will serve to bring out the change that followed the transfer of control to the forest service in the midst of the fiscal year 1905:

- A large excess each year in the amount of timber sold over that cut and paid for;
- Nine times as much timber sold at the end of the four-year period as at the beginning and three times as much cut;
- A much higher price obtained per thousand board-feet at the end of the period than at the beginning. Each of these matters calls for comment. The sales are of stumpage only; the government does no logging on its own account.

- More timber is sold each year than is cut and paid for, because many of the sales extend over several years. With increasing sales the amount sold each year for future removal has exceeded the amount to be removed during that year under sales of earlier years. Large sales covering a term of years are made because the national forests contain much overmature timber, which needs removal, but which is frequently too inaccessible to be saleable in small amounts. To prevent speculation the time allowed for cutting is never more than five years and cutting must begin at once and be continued steadily.
- The volume of sales has increased rapidly because much 1 The United States fiscal year ends June 30 and receives its designation from the calendar year in which it terminates. Thus, the fiscal year 1898 is the year July 1, 1897 - June 30, 1898.

Administration transferred to Bureau of Forestry, February I, 1905. Full utilization of all merchantable material, care of young growth in felling and logging and the piling of brush, to be subsequently burned by the forest officers if burning is necessary, are among these conditions.

Timber to be cut must first be marked by the forest officers. Sales of more than $100 in value are made only after public advertisement.

Only the simplest forms of silviculture have as yet been introduced. The vast area of the national forests, the comparatively sparse population of the West, the rough and broken character of the forests themselves and the newness of the problems which their management presents, make the general application of intensive methods for the present impracticable.

Natural reproduction is secured. The selection system is most used, often under the rough and ready method of an approximate diameter limit, with the reservation of seed trees where needed. The tendency, however, is strongly towards a more flexible and effective application of the selection principle, as a better trained field force is developed and as market conditions improve.

One conspicuous achievement was the reduction of loss by fires on the national forests. During the unusually dry season of 1905 there were only eight fires of any importance and the area burned over amounted only to about 16 of 1% of the total area. In 1906 about 12 of 1% was burned. This was accomplished by efficient patrol, co-operation of the public and by preventive measures, such as piling and burning the brush on cut-over areas.

Since the beginning of 1906 the largest source of income from the national forests was their use for grazing. Stock-raising is one of the most important industries of the West. Formerly cattle and sheep grazed freely on all parts of the public domain. In the early days of the national forests the wisdom of permitting any grazing at all upon them was sharply questioned.

Unrestricted grazing had led to friction between individuals, the deterioration of much of the range through overstocking and serious injury to the forests and stream flow. The forests of the West, however, are largely of open growth and contain many grassy parks, the results of old fires and many high mountain meadows.

Under proper regulations the grass and other forage plants which they produce in great quantity can be used without detriment to the forests themselves and with great benefit to the stock industry, which often can find summer pasturage nowhere else.

Except in southern California grazing is now permitted on all national forests unless the watersheds furnish water for domestic use; but the time of entering and leaving, the number of head to be grazed by each applicant and the part of the range to be occupied are carefully prescribed. Planted areas and cut-over areas are closed to stock until the young growth is safe from harm and goats are allowed only in the brushland of the foothills.

The results of regulation, in addition to the protection of forest growth and streams, are the prevention of disputes, improved range, better stock, stable conditions in the stock industry and the best use of the range in the interest of progress and development. The first right to graze stock on the forests is given to residents, small owners and those who have used the range before. Thus the crowding out of the weaker by the stronger and of the settler by the roving outsider has been stopped. In 1906 the forest service began to impose a moderate charge for the use of the national forest range.

The following statement shows the amount of stock grazed on the national forests 1904-09 and the receipts for the grazing charge: - A work of enormous magnitude which has now begun is planting on the national forests. At present, with low stumpage prices and incomplete utilization of forest products, clear cutting with subsequent planting is not practicable.

There are, however, many million acres of denuded land within the national forests which require planting. Such planting is still confined chiefly to watersheds which supply cities and towns with water. The first planting was done in 1892, in California. Since then similar work has been done on city watersheds in Colorado, Utah, Idaho and New Mexico. Other plantations are in the Black Hills national forest, where large areas of cut-over and burned-over land are entirely without seed trees and in the sandhill region of Nebraska. Up to 1908 about 2,000,000 seedlings had been planted, on over 2000 acres - a small beginning, but the work was entirely new and presented many hard problems.

The nursery operations of the forest service are concentrated at seven stations, located in southern California, Nebraska, Colorado, New Mexico Utah and Idaho, where stock is raised for local planting and for shipment elsewhere.

These nurseries are small. Their annual productive capacity is between 8,000,000 and seedlings. Each nursery is practically an experimental forest-planting station, at which a large variety of species are grown and various methods are tried.

The organization of the administrative work of the national forests is by single forests. On the 1st of January 1908 the total number of forests was 165 with a total area of 162,023,190 acres. In charge of each forest is a forest supervisor. Under the supervisors are forest rangers and forest guards, whose duties include patrol, marking timber and scaling logs, enforcing the regulations and conducting some of the minor business arising from the use of the forests. Guards are temporary employes; rangers are employed by the year. The supervisors report directly to and receive instructions from the central office at Washington.

In this office there are four branches - operation, grazing, silviculture and products - each of which directs that part of the work which belongs to it, dealing directly with the supervisor.

For inspection purposes, however, the forests are separated into six districts, in each of which is located a chief inspector with a corps of assistants. The inspectors are without administrative authority, but assist by their counsel the supervisors and through inspection reports keep the Washington office informed of the condition of all lines of administrative work in progress. Administrative officers alternate frequently between field and office duties.

The number of forest officers in the several grades on the 1st of January 1908 were: 6 chief inspectors, 26 inspectors, 106 forest supervisors, 41 deputy forest supervisors, 820 forest rangers and 283 forest guards. The total number of employes of the forest service on the same date, including the clerical force, was 2034.

Besides the administration of the national forests, the forest service conducts general investigations, carries on an extensive educational work and co-operates with private owners who contemplate forest management upon their own tracts. This last work is undertaken because of the need of bringing forestry into practice, the lack of trained foresters outside of the employ of the government and the lack of information as to how to apply forestry and what returns may be obtained. Co-operation takes the form of advice upon the ground and, on occasion, of the making of working plans. The educational work of the service is performed chiefly through publications, the purpose of which is to spread very widely a knowledge of the importance of forestry to the nation and of the principles upon which its practice rests. The investigations which the service conducts extend from studies. of the natural distribution and classification of American forests. and of their varied silvicultural problems to statistics of lumber production and laboratory researches which bear upon the economical utilization of forest products. As examples of these researches may be mentioned tests of the strength of timber, studies of the preservative treatment of wood for various uses, wood-pulp investigations and studies in wood chemistry.

THE ASSESSMENT FRAMEWORK AND FOREST POLICY

Reviewing the various reports on the issue of forest policy has provided an insight into the process by which decisions are made. To summarise the inadequacies that have been identified in the respective assessment frameworks and processes.

Assessment Frameworks and Processes

There are several assessment frameworks used in Australia for forest decision-making. Most of the reports have only considered assessment frameworks in general terms, without specific reference to real examples. In order to evaluate these assessment frameworks, the report provides a 'checklist'. The 'checklist' provides a comprehensive series of questions which any worthwhile assessment framework should be able to answer. The 'checklist' approach facilitates identification of the broad strengths and weaknesses of existing or proposed assessment frameworks

A common theme in all reports (both Australian and international) was an underlying philosophical assumption that a 'technical', non-political choice process was possible, and to some extent desirable. This process was not challenged explicitly in any report, despite the obvious claim that the political arena is the desired place for settling value disputes (indeed it could be argued that this is the raison d'etre of politics). For the moment, however, this discussion accepts the premise that a 'technical' or 'black box' approach is feasible for value choices. So as to gain an understanding of international experience, the report has sought to review processes suggested by the OECD and the World Bank. These internationally-recommended approaches provide a benchmark for achieving 'world best practice' in this field. The report has also examined the type of system deployed by New Zealand under the auspices of the Resource Management Act.

The international reports reflect a worldwide concern over the assessment of environmental values. While it is now recognised that environmental values must be incorporated, it is not known how this should be achieved. The World Bank view is not unusual:

Externalities are thus clearly troublesome, and there is no altogether satisfactory way to deal with them. This is no reason simply to ignore them, however; an attempt should always be made to identify them and, if they appear significant, to measure them. This accurate, but unhelpful commentary is largely due to a neglect of environmental factors within economic analysis. This was noted by two internationally renowned economists in 1988, with respect to the field of development economics:

Environmental resources appear in this literature about as frequently as rain falls on the Sahara. They also noted, as recently as 1988, that one of the world's best known texts on development economics... simply has no discussion of environmental resources and their possible bearing on the development processes.' This is symptomatic of neglect within the broad field of economic analysis. The OECD (1992) remarked that conventional cost benefit analysis would not include a number of environmental effects ...because of the lack of market price data or easily accessible direct proxy price data.'

Given this neglect the OECD report (1986) on the public management of forests, must be regarded as a landmark publication amongst the reports reviewed in this study. Overall, given that all economic activity centres on the idea of 'development', the general lack of incorporation of environmental issues is a serious omission. As a result the weaknesses identified in Australian assessment processes are not unusual when compared with 'world best practice'-though this does not make such deficiencies acceptable according to economic theory. Neo-classical economic theory would state quite clearly that on a worldwide basis, as well as locally, such an omission must be regarded as a threat to the economic well-being of society.

Existing Assessment Frameworks

Australia, with its federal political structure, has many different means by which choices over values are made. Rather than describe this diverse array of assessment frameworks, this book will provide a thumbnail sketch of the key features of assessment frameworks and processes. Along with the international examples cited, seem to point to three key characteristics of concern in any assessment: 'Who makes the decisions' is significant because it determines whose values are brought to the process. The economics profession has long argued that institutions tend to be captured by the values of specific groups, and that to expect any public institution to be value-free and working purely in the public interest is naive. From the perspective of assessment processes, it is important that advocates of differing values are given an equal voice in the process of analysis and decision-making. The RAC Forest and Timber Inquiry report took some interest in the Victorian model which involved the Land

Conservation Council in decisions about selection of forest logging coupes. The AHC-CALM model of joint survey work also received some favourable attention, though its joint deliberations were designed to identify values rather than make decisions. That no other systems were singled out for favourable comment, in the various reports reviewed in this study, seems to indicate that other processes may be lacking an adequate contribution from representatives of other values.

It is not unfeasible that an appropriately structured institutional process could make optimising decisions-given that the right information on values was available. The information provided, as outlined by the thirteen reports reviewed in this process, is inadequate for the task of making choices over values. The AHC-CALM model received some favourable comments for its process from the ESD and RAC reports Certainly, its assessment method is systematic and thorough. Its prime weakness is that it does not consider, in its own words, 'social values'. 'Social values' as defined by the AHC, appear to cover that group of values described by economists as existence value. As a result, it only acknowledges some of the values considered important in a total economic assessment. That it was commended as a model for others suggests that even analysis of 'non-social' values is deficient in existing assessment frameworks and processes.

Other reports such as the Victorian Auditor-General's report and the NSW Public Accounts Committee confirm the impression that only poor data is available on values. Financial values are also a significant part of the analysis. A significant range of reports located deficiencies or made recommendations that pointed to opportunities to improve the gathering of financial data. The deficiencies in this regard extended to the data on both financial values generated by forestry and financial values generated by other activities. The existing assessment framework and processes were not being supplied with the relevant data because the necessary mechanisms such as appropriate accounting systems are not adequate.

Compare Different Values

Assessment frameworks must make choices between values. Few technical models were offered on how such choices should be made,outside the political process. The Victorian Auditor-General's report noted that the previous system in that State had used a committee of relevant officials, backed up by a requirement to find options, with a final 'safety valve' being provided by there being recourse to the Minister. No doubt similar methods are used elsewhere, though, what is required is a method that considers all values at the point of decision with appropriate systemic checks and balances.

The AHC-CALM model and the process offered by the RAC Forest and Timber inquiry were the closest Australian examples of technical approaches which are beginning to seek full consideration of all values. The OECD (1986) report is a benchmark in this field. Only the process implied in the RAC report comes close to this analysis. It is assumed that if there were another system, the RAC inquiry would have located it. The existing assessment frameworks, obviously, can be said to rely on the judgements of those who take the decisions and the information that they have been supplied with. There must be serious doubts that these existing frameworks have been properly structured. A assessment framework and process needs to specify who decides, with what value information and by what intervalue comparative techniques.

Inadequacies in Pricing/financial Valuation

The significance of non-financial values to the calculation of financial values is considerable. The forest agencies and all the reports reviewed in this study have adopted a compartmentalised approach to pricing. The compartmentalised approach assumes that assessment of choices between values occurred prior to the pricing of timber. If non-financial values are improperly assessed, then the resulting effect is that financial values will also be inappropriately founded. In this sense a compartmentalised approach to pricing is a practical manner of settling commercial pricing issues. In simple terms, if non-financial values are being compromised it is due to the lack of adequate assessment prior to the commercial pricing process.

SOCIAL PARTICIPATION AND FOREST POLICY

Decentralization in Bolivia was motivated by increasing pressure from civic committees seeking to have greater control over natural resources, the general trend towards decentralization in neighbouring countries and the prominence of decentralization in donors' agendas. Its implementation has been affected by social participation dynamics, and the new forestry regulations have had implications for democratization of forest resources access.

ANTECEDENTS OF DECENTRALIZATION

Bolivia has three levels of government: central government; departmental government or prefecture, whose main authority (the *prefecto*) is appointed by the president; and municipal government, in which democratically elected municipal councils elect the mayor. Before decentralization most decisions were made at the central level. The municipalities had limited resources and little influence in policy-related decisions, even those directly affecting the development of their municipal jurisdictions. The Bolivian lowland forest area was largely marginalized from the political centre in La Paz, where most political decisions were made. In the early 1960s that region was progressively integrated into the national economy through the expansion of natural gas extraction, agriculture and logging. Those factors fostered the growth of a regional elite, expanded the contribution of that region to the national income and increased its influence on development programmes. In the 1970s, the government took its first step towards decentralization by establishing corporations for regional development. This was closer to an administrative attempt to transfer some investment decisions to the departmental level but the central government still appointed the presidents of such entities, and most of the decisions were negotiated at higher levels. Civic committees (groups of local social organizations) had been seeking greater access to forest revenues and greater participation in the formulation of forest policies since the mid-1970s.

In the 1980s, legislation was approved establishing the collection of a forest fee of 11 per cent, to be used in regional development projects. (In 1993, further legislation was passed stating that forest companies would pay 80 per cent of their taxes directly in the areas where the resources originated, but the mechanism was hard to implement in practice.) By the late 1980s, the national forestry service was deconcentrated to local branches. This move towards deconcentration, however, did not make the forestry service a more efficient and less corrupt institution. Municipal authorities, meanwhile, continued to respond to the leaders of their respective political parties rather than to their constituents.

Decentralization was prompted through approval of Popular Participation Law (No. 1551) and Administrative Decentralization Law (No. 1654), both passed by Congress in 1994. The first altered the responsibilities of the municipal governments, while the second modified the responsibilities of prefectures or departmental governments. In 1996 a new Agrarian Reform Law was issued, as well as a new Forestry Law. The Agrarian Reform Law's objective was to define the legal basis for a system of titling and land regularization, and to redefine the conditions of access to rural property. The Forestry Law attempted to redefine the conditions for obtaining and maintaining forest rights. Both affected the way in which landholders and forest users can access and maintain their rights for forest resources use.

PARTICIPATION IN INVESTMENT DECISIONS

The Popular Participation Law expanded the municipal government's jurisdiction beyond the urban centers to the whole rural area within the municipal borders. It made municipalities responsible for local schools, health facilities, roads maintenance and water systems. To finance these new responsibilities, the central government allocated 20 per cent of the national budget to the municipal governments, to be distributed among municipalities in proportion to their populations. Both rural and urban property taxes were earmarked for the municipal governments, who now administer their collection. The Popular Participation Law has strengthened municipal governments and made them more democratic. Rural populations – mainly smallholders and indigenous people – have gained the right to participate in municipal elections and run for the municipal councils. Nonetheless, national political parties still appoint individual candidates, allowing political parties to maintain their control over local political agendas, and to reproduce a system of political patronage with local leaders. Local candidates who want to run for office have to negotiate their agendas within the priorities of the political parties, and if they are elected they are accountable to those parties.

Furthermore, the law sought to introduce community control over municipal governments by local social organizations (ie local farmer organizations, neighbourhood committees and indigenous groups) and community-based vigilance committees. Nevertheless, these committees' representatives are exposed to continuous pressures from political parties.

Rules and Regulations for Forest Use

The Forestry Law of 1996 defined a set of regulations for forest use, somewhat differentiated according to forest user, under the premise that sustainable forest management is feasible under the right practices. It established a new system for monitoring forest management, enforcement and sanctions to illegal logging, as well as introducing some market-oriented regulations and taxes to discourage unsustainable forestry operations. The goal is to achieve sustainability of forest management through progressive incorporation of less valuable timber species and the application of extraction techniques that promote natural regeneration. Furthermore, the law seeks to create clear rights over forest resources, thereby encouraging investments in forest management; and to eliminate forest crime and illegal logging, as well as set technical criteria for forest management. The public institutional system of the forest sector was substantially altered. The Ministry of Sustainable Development and Planning is the ruling entity, the Forestry Superintendence is the regulation entity, and the National Forestry Development Fund is the financial entity.

Noncommercial forest uses do not require authorization, and a forest management plan is an essential requirement for all types of commercial forest activities. Hence, forest concessionaires as well as private landholders are obligated to design management plans as an instrument to regulate commercial logging activities, including forest inventories and mapping.

Forest management plans have to comply with many technical requirements. Forest management, when based on selective management, must respect a minimum cycle of 20 years between logging operations on the same area, and a minimum cut diameter must be respected. Furthermore, annual operations plans are required. The regime to be applied to nontimber forest products is similar.

Democratizing Access to Forest Resources

The new forestry regulations included two provisions that have had some impact in democratizing access to forest resources. The first refers to the exclusive right of indigenous peoples to use the forest resources within their indigenous territories, recognized legally by the Agrarian Reform Law. According to this law, indigenous claims are considered titled after completion of a process of land regularization. Currently, a total of 19 million ha have been claimed for titling as indigenous territories, but it is not known how much will ultimately be effectively titled in this way. Stocks (1999) estimated 5 million ha to be the area with commercial logging potential in areas claimed as indigenous territories. A provision of the Forestry Law states that local forest user groups can benefit from forest concessions within areas declared as municipal forest reserves, which represent up to 20 per cent of public forest within each municipal jurisdiction. Local forest users can be granted a forest concession if they are recognized as a local user association by the Ministry of Sustainable Development and Planning. This mechanism was conceived as a way to formalize the access for local forest users or small-scale loggers who, because they had no legal right to access forest resources, were previously conducting forestry operations informally.

POWERS TRANSFERRED TO MUNICIPAL GOVERNMENTS

According to Ribot (2001), outcomes of decentralization depend on the type of powers that are transferred to lower levels and on who receives such powers. In Bolivia, as well as in most countries of Latin America, decentralization has followed a top-down format, and municipalities have been the main recipients of authority transferred from the central level.

A Top-Down Model of Delegation

Although the Popular Participation Law did not grant municipal governments any new explicit function related to natural resources management, it motivated some municipal governments, as a result of their larger political authority, to become involved in natural resources issues and thereby capture part of the benefits. Because mayors became relatively more powerful, the central government and the international donors began to consider them more seriously as partners in environmental projects. The increasing political power of municipalities led the government to consider them for dealing with some problematic issues, such as social monitoring of illegal logging and formalizing forest rights for small-scale loggers and other local users. These were included in the Forestry Law of 1996.

The current powers and functions are divided as follows:

- Formulate forest policies, strategies and norms;
- Determine land classification and evaluate forest potential;

- Prepare demarcation of concession areas for timber companies and local groups;
- Set prices for concession fees and volume-based taxes;
- Promote research, extension and education; and
- Solicit technical assistance and funding for forestry projects.

Forest superintendence (SF):

- Supervise technical compliance with the forestry regime;
- Grant management rights to eligible forest users;
- Approve forest management plans for different forest rights;
- Enforce forest regulations and sanction illegal forest users;
- Issue concessions, authorizations and logging permits;
- Request external forest audits of forest operations; and
- Collect concession fees and volume-based taxes and distribute them.

Prefectures:

- Formulate forest development departmental plans;
- Develop forest research and extension programmes;
- Promote programmes of rehabilitation of degraded forest systems; and
- Develop programmes for strengthening municipal forestry units' institutional capacities.

Municipal governments:

- Propose the delimitation of municipal forest reserves up to 20% of available public forest;
- Protect and conserve the reserve areas until they are conceded to local user associations;
- Inspect and control all forest activities within their territorial jurisdiction;
- Report violations of forest regulations to the Forestry Superintendence;
- Provide support to local forest users in implementing their management plans;
- Establish the registry of forest resources in their jurisdiction;
- Develop soil use plans corresponding to the departmental use plans;
- Organize training events for local forest users; and
- Facilitate and promote social participation in local forest development.

To carry out their new responsibilities, municipal governments are expected to create municipal forestry units. Municipalities can form consortiums with other municipalities to create such units. In theory, the entire system should be entirely financed with the revenues coming from both concession and clearcutting fees. Prefectures receive 35 per cent of the concession fees and 25 per cent of the fees charged for clearcutting operations. Municipal governments get 25 per cent of both types of fees. The National Forestry

Development Fund receives 10 per cent of the concession fees and 50 per cent of the clearcutting fees. SF gets 30 per cent of the concession fees. A reduction of forest concession fees in March 2003 has diminished the financial resources going to prefectures and municipalities.

Restricted Powers for Municipalities

Municipal governments must comply with national regulations regarding property regimes and forest regulations. Hence, their autonomy to make decisions about natural resources depends largely on decisions made at the central level. The legislation thus regards municipal officials as rule followers rather than rule makers, and as implementing agencies for policies defined at the central level. In short, the new institutional forest system has not delegated important responsibilities to municipalities and has therefore not led to dramatic changes. The central level reserves for itself major decisions regarding allocation of forest resources rights, granting forest concessions, approving forest use regulations, and collecting taxes from forest resources use. Municipalities have little to say in these areas, and all informal actions of municipalities for collecting forest taxes and controlling timber transit are considered illegal.

Although municipal governments are empowered to control forest crime, promote community forestry and take other forest-related actions, their power to make autonomous decisions regarding the allocation and use of forest resources is restricted. For instance, they can decide how to allocate the forest resources within the municipal reserves but cannot decide about the size of such reserves. This reflects the tension between the central level, which defends the strong role of the national forestry service, and those who support a more active and autonomous role for municipal governments.

Limited Autonomy for Local Forest Users

The situation of indigenous groups has improved. Indigenous communities whose forest resources were subject to encroachment from illegal loggers now have exclusive access to resources within their territories, though problems persist. However, before indigenous people can take commercial advantage of forest resources, they have to develop forest management plans according to Forest Superintendence regulations, and they have little scope to adapt such norms to their own management practices (BOLFOR, MDSMA, and Superintendencia Forestal, 1997).

And even though specific regulations for forest management in indigenous areas have been approved, local knowledge has been ignored in much of the forest policy. Enforcement of indigenous property rights depends on some institutional arrangements made at the departmental and regional levels, over which indigenous people have little influence. This issue is linked to a more complex bureaucratic process of land use planning and titling with poor overall outcomes. The other social actors who have benefited from access to forest resources – small-scale timber extractors and other local forest users – also have no autonomy to make decisions about the way in which they use the resources, and all of them have to comply with forestry regulations to maintain access to public forest resources through forest concession systems.

The Forest Superintendence has instituted command-and-control mechanisms to enforce the implementation of what it considers good forest management practices among local forest user groups and small landholders. Those practices, though appropriate for large-scale forest concessionaires, act as barriers for small-scale forest users.

Implementation of Decentralization Policies

There are always failures in the implementation of policies, and decentralization is no exception. In Bolivia, transferring responsibilities to municipalities meant building local capacities in municipal governments, as well as interacting with actors who had disparate incentives and interests to use the forest within the new institutional context.

Building Institutional Capacities

Municipalities with forest resources began to receive their shares of forest taxes in 1997, when they started to set up municipal forestry units. The resources transferred to municipalities have tended to decrease because forest concessionaires did not comply with the forest fees payments. In 2002, a new system formally reduced the amount of taxes collected by the state from forest operations to almost half of what was originally expected, formalizing the reduction of transfers from the central government to municipalities. By the end of 2001, almost all municipalities with forest resources (about 109) had created their municipal forestry units, and each had at least one forest or agricultural technician. In municipalities with difficult access and little population, the forest units are the only local providers of technical services, and governments tend to value their role in local planning. The main limitations are the staffs' lack of technical skills and inadequate training in social issues, such as conflict resolution.

Building technical capacities in municipal forestry units constituted an important step forward in relation to the past. Nevertheless, progress is not uniform and varies with the financial support the units receive from the municipalities. That in turn depends on the amount collected in forest fees and what proportion of it goes to supporting operational budgets. A significant portion of municipalities consider it enough to provide minimum resources to the municipal forestry units and spend the rest, if any, on other activities. It is worth mentioning that local government priorities are providing social services and infrastructure, and they have little motivation to support productive projects, though this is changing.

Incentives for the Actors

The new institutional system resulting from both decentralization and the shift in forest policies has modified the incentives for municipalities and forest stakeholders to continue doing what they were used to doing, or to adapt their social and financial strategies to the new conditions. The main incentive for municipalities to engage with the new decentralized order of things was the likelihood of getting a share of the forest taxes, as well as the possibility of administering the municipal forest reserves. Municipalities also are interested in penalizing illegal clearcutting due to the fact that they get a portion of the fines, but this does not happen in relation to illegal logging. Furthermore, municipal governments have been interested in being active players on forest issues in the cases where local forest users have political influence on local decisionmaking, or in cases where forest-dependent people represent important votes to keep them in office.

The incentives of local forest users to engage in the process are diverse. Indigenous people have received important benefits from forest policy reform and decentralization. They have gained rights to make exclusive use of their forest resources within their territories, and have the chance to expand their influence to participate in municipal decision making. In this context, forest management might increase the social legitimacy of indigenous people's claims to land.

In turn, local small-scale timber extractors' main incentive has been to get formal access to forest resources through the forest concession system, and to benefit more from formal markets. Furthermore, this group should benefit from the technical assistance provided by the municipal forestry units, which are receiving resources from the forest taxes to support such activities. Both groups have benefited from several forest projects (eg BOLFOR) and NGOs with an interest in supporting community forestry.

In the short run, the main losers under decentralization were absentee forest concessionaires, who must now negotiate with the forestry service. Furthermore, a forest tax has been imposed, and they have had to acknowledge indigenous people's demands. The main incentive for forest concessionaires to adapt to the new conditions was their interest in keeping their forest areas, and an implicit commitment from the central government to help them through a difficult financial situation originating in the timber crisis in regional markets. Municipal governments reacted in contradictory ways regarding forest concessions. Although some are critical of concessions' activities within their jurisdictions, others consider them sources of economic growth and employment.

Homogeneous Solutions for Heterogeneous Municipalities

The Bolivian model of decentralization does not account for regional variation, and its design assumes that all municipal governments will react uniformly to the challenges arising from the new conditions. Reality proved rather different. Implementation of decentralization has had to face three issues: uneven distribution of resources to municipalities, the minimal availability of public forest declared as municipal forest reserves, and the hetereogeneous interests of municipalities regarding forest-related activities.

Uneven distribution and allocation of financial resources. The financial resources allocated by the Forestry Superintendence to municipalities have varied. From 1997 to 1999, only 30 municipalities benefited, with more than 80 per cent of the total resources transferred to them. In the richer municipalities, not all income from forest taxes is spent on forest-related activities; some is diverted to other sectors. The poorest municipalities likewise have urgent demands that are outside the forestry sector.

Minimal municipal forest reserves. Even though the law specified that up to 20 per cent of public forest would be declared municipal forest reserves, in practice such areas were not available in all municipalities – in some cases because of the existence of overlapping claims over public forest, in others because public lands were maintained as forest concessions. In some municipalities there were no areas to declare as municipal reserves to allocate as concessions to small-scale loggers. By early 2001, from a total of 2.4 million ha demanded by municipalities as municipal reserves, only 681,000 ha had been designated.

Municipalities' priorities. Only a few municipal forestry units have accomplished the functions they have been granted. Municipalities prioritize their investment in response to social pressures or to local authorities' political motivations. The level of municipal engagement is also related to municipalities' incentives to respond to their constituents and be accountable to higher authorities. For instance, whereas some municipalities are interested in allocating resources to establish a system of forest concessions within municipal reserves, others choose to control illegal clearcutting or make forest management viable for small farmers.

Uneven Outcomes

Transferring responsibilities to municipalities has modified the political and institutional arrangements for forest management, with implications for both people and forests.

Participation in Local Politics

In many lowland municipalities, small farmers, indigenous people and small-scale loggers have been elected to office for the first time. Indigenous groups have been able to obtain political support from the municipal councils to reinforce their land claims, small-scale loggers have obtained support to negotiate temporary logging authorizations, and small farmers have obtained support favouring their efforts to modify land-use and forest regulations. Thus, municipal governments have contributed to political support for some local actors' claims on resources. This is also the case where these groups have strong organizations that can influence the municipality's decisions or where they represent the majority of voters. In these cases, municipal governments may amplify the demands of social actors. In other cases, however, transferring responsibilities and resources to municipalities has reinforced the power of preexisting local elites – elite capture, particularly in municipalities of northern Bolivia, and where cattle ranchers and timber companies are highly influential in local politics. These local elites have, in some cases, influenced municipal governments to build local alliances against indigenous land claims and strengthened their power over use of resources. The local elites of these municipalities often benefit from activities based on mining of natural resources and promote concentration of benefits in a few hands. Decentralization, in these cases, tends to produce undesired results in social equity and forest conservation.

Where the social composition of local government is more complex, local elites have to negotiate with small farmers and indigenous people. It is not unusual to find alliances built among the different social groups that support certain development agendas, such as infrastructure development and basic social services. Those social agreements are, however, difficult to arrange for natural resources management, though in some cases alliances have been established to protect conservation areas against encroachment of foreign timber and mining companies (ie municipalities of Rurrenabaque and San Ignacio de Velasco).

Much of the impact of decentralization on reconfiguring social participation in local politics depends on the local political economy and on the social capital of marginalized groups. In other words, the more democratic municipalities have traditionally had more equitable access to resources. Those in which local elites have imposed their interests are affected by the extent to which local groups have been able to build social capital.

FORMS OF CULTURALLY-DERIVED EVIDENCE IN BRITISH COLUMBIA

Historical evidence, or information collected was organized into four broadly defined categories, including: (1) aboriginal history, (2) written records, (3) oral history, and (4) maps and photographs. Although distinct categories were established for purposes of facilitating data analysis, sources were by no means mutually exclusive, and considerable overlap occurs between each source. For example, although maps and written documents are distinct categories, most of the maps used in this study were found within written documents. Or, although aboriginal history and oral history are separate categories, a key form of aboriginal history includes oral history.

ABORIGINAL HISTORY

Aboriginal land use and management practices have profoundly influenced ecosystems at multiple levels. Both plants and animals were harvested and managed sustainably. The knowledge gained from the close ties aboriginal peoples have to the land is termed Traditional Ecological Knowledge (TEK). Many authors consider TEK information complementary to scientific and other forms of information. Ecological restoration may not be complete if it fails to consider the rich knowledge, traditions and practices of aboriginal cultures. TEK can provide evidence of the biological richness of historical ecosystems, and provide models for sustainable resource use. According to Anderson (2001) aboriginal history may be further classified as (a) oral histories and participant observations with aboriginals, (b) ethnographic literature, (c) museum artifacts, (d) ecological field experiments, and (e) plant and animal remains in an archaeological context. In this work, several sources of ethnographic literature are used to support ecological restoration on the JPRF.

The JPRF is situated in the traditional territory of the Tl'azt'en Nation, located north of the village of Fort St. James in the Fraser River watershed. Unlike many aboriginal communities which have not yet documented their experience and knowledge of the land, Tl'azt'en Nation have compiled an extensive body of oral and cartographic information related to their cultural experience and knowledge.

Tl'azt'en Nation granted access to several information sources housed in their Natural Resources and Administration offices, located in the community of Tache, British Columbia. The first source was a collection interviews from Dr. Annie Booth, examining the Tl'azt'en Nation's Tree Farm License (TFL) 42. This source provided a significant body of information including interviews with elders, Band Council members, forest workers, individuals linked with TFL 42 and other community members. While the focus of this research was on TFL 42, discussions included traditional land management in Tl'azt'en traditional territory, and industrial land use history within the JPRF. The second information source included summaries of interview and focus group transcripts conducted with Tl'azt'en elders between 1978 and 1995, which concentrated on traditional life in Tl'azt'en territory prior to 1950.

Written Records

Written records are useful for evaluating change in species composition, age distribution and landscape structure over time. An extensive body of written records exists documenting past land use practices and natural history of British

Columbia, and written records that support ecological restoration on the JPRF include both manuscripts and government records. Manuscripts include unpublished, non-government records generated by individuals, families, organizations or businesses. Access was granted by experiential knowledge holders to several unpublished reports regarding land use practices in and around Fort St. James.

Three permanent sawmills are known to have existed on the JPRF, and access was granted to original documents produced by Industrial Forestry Services (IFS), a consulting company with a long operating history in central-interior British Columbia. These records relate to one licensee who operated on the JPRF and surrounding areas. The first report focused on the timber supply for operations based on Tezzeron Lake, which specifies a tentative harvesting plan and the timber potential throughout the JPRF. The second report includes a cruise and partial logging plan for a timber sale on the JPRF. This report contains a detailed description of stand conditions prior to timber harvesting and provides supplemental information regarding land use practices. Next, an inventory of the area encompassing the Tezzeron Lake operation summarizes the volume of merchantable trees in the area where Park Brothers Ltd. was operating, including the JPRF. The fourth report details the quality and quantity of mature forest available for harvest throughout the Stuart Lake Public Sustained Yield Unit (PSYU) and operations which persisted within the Stuart Lake PSYU in the mid 1960s Industrial Forestry Service (IFS). 1965b. Park Brothers Limited: inventory of Tezzeron Lake operating unit. Industrial Forestry Services, Prince George, BC.

Unpublished Report

Finally, an evaluation of waste details trees that were cut but not hauled to mill sites on two timber sales on the JPRF. Also, access was granted to several unpublished reports produced by an experiential knowledge holder which describe land use history in the Prince George and Fort St. James areas, and silvicultural systems employed in these areas. Unfortunately, the value of written records for understanding forest history and for supporting ecological restoration in British Columbia is generally not well known or appreciated. For example, the owner of one lumbering company which operated on the JPRF disposed of all records relating to past timber harvesting one year prior to this research. "Just a year ago I took out, I had all the old timber sales contracts, and maps and that stuff just exactly a year ago I took it out and put it into a bonfire" – Owner/operator Cassiar Forest Products, 2001

Government records include documents created by provincial government departments or ministries. Those records that are particularly relevant to ecological restoration include records that relate to land and resource management. These include timber records (*e.g.* reconnaissance, cruising and sales), Department of Agriculture records, lot registers, water rights records and park records. Government records contain information about government policy regarding land management, the types of vegetation in an area, agricultural practices, land settlement and division, and decisions surrounding resource use and conservation. Although records relating to mineral exploration associated with the Pinchi Lake Mercury Mine are believed to exist, it was not possible to obtain these records for use in this research.

Two key forms of government records were used in this study. First, access was granted to an inventory of abandoned sawmill sites in the area, as prepared by the Fort St. James Forest District. This inventory contains maps and photographs of abandoned sawmill sites and was useful for understanding how sawmills were distributed across the JPRF.

The second form of government records used in this research includes timber sale files. Timber sale files were created to document timber allocation and harvesting activities from 1912 through the 1970s in British Columbia. These files vary in technical quality, but generally contain information on land cover, management practices, ownership, physical features and timber yield within the bounds of the timber sale. Timber sale files present an opportunity to extract detailed information about land use and landscape structure at relatively small spatial scales. Eighty-six timber sale files were uncovered for the JPRF, spanning a thirty-five year period (1940 to 1975). Timber sale files were accessed through the British Columbia Forest Service, Inventories Division, in Victoria. The files were organized chronologically and an index summarizing the files was compiled. A summary of the documents contained within each file was created and included in the index. Additionally, an ecological summary based on any ecologically relevant information contained in the file was written for each timber sale and included in the index.

Oral Histories

Many users of the land have considerable knowledge about land use history and historical ecosystem conditions, but few individuals have recorded this information. When working within the bounds of an oral history, however, people are often able to recall this information. An oral history is a structured conversation between an interviewer, who is pursuing a specific range of topics, and an interviewee, who holds the information related to those topics.

To date, oral sources have most often been used to develop an understanding of land use history, and less often been used to interpret historical ecological conditions. In the context of this research, oral histories include new interviews conducted to support this study. Candidate interviewees were selected to represent a diversity of perspectives related to past land management practices and historical ecology on the JPRF. Additional interviewees were identified using the snowball sampling method, where the interviewee is asked about other potential interviewees. In total, ten individuals were interviewed using a semi-structured interview guide to organize the interview-discussion.

When individuals agreed to participate, they were sent a detailed letter explaining the project, rationale for their selection, and the ethical considerations related to the project. After the information package had been received and read, interviewees were contacted again to arrange an interview and clarify any of the interviewees' questions. Overstory and understory vegetation, roads, riparian areas, fire, insects and diseases were important forest condition themes identified from the literature, and interview questions were designed to generate discussion about these specific forest characteristics. Questions were also asked about land use practices and land use history to generate discussion about changes in forest condition. Interviewees were requested to disclose demographic information to determine which particular geographic areas they were familiar, and to what time period their information could be applied.

Interviews were transcribed in full following the interview. When possible, both the interviewer and interviewee edited the transcripts. Interviewees were given the opportunity to omit any portion of the conversation and to ensure names and facts were correct to the best of their knowledge. The edited versions of the transcripts were used for analysis. Transcripts were analyzed to summarize information related to criteria described above, and ecological summaries from each interview were combined to identify similarities and differences in information. Interviewees were also grouped according to geographic expertise and time period of experience.

Maps and Photographs

Maps and photographs present the most obvious source of information regarding past landscape conditions where details are spatially comparable over time. This comparison is possible because maps and photographs have natural landmarks such as hills, lakes and streams that are very similar to landscape features that exist today. Details such as vegetation types may be interpreted, if the researcher has an understanding of the map-making standards at the time the map was produced. Maps used in this research generally accompanied written records, such as timber sale files, and in these cases it was possible to extrapolate the information from written records to specific areas on the map.

Direct evidence of the past can be found in photographs, and comparisons may be made between photographs taken of the same area, at a similar scale, over different time periods. Photographs are a selective record of the past, as attractive scenes or culturally significant sites have a higher likelihood of being photographed.

Also, photographs represent the three-dimensional world in only two dimensions. Several still photographs have been uncovered which depict the natural environment and land management practices on the JPRF. The photographs taken of natural areas, however, were not specific enough to permit repeat photography of the area. The photographs documenting management activities offer visual insight into the activities and practices which caused changes on the landscape, such as sawmill equipment and harvesting practices.

Changes in landscapes are also visible through vertical aerial photography; however, in some cases aerial photographs may postdate the desired time period for restoration activities. Changes in vegetation can be digitally analyzed over time using aerial photography, digital image processing and geographic information systems (GIS). Aerial photographs became generally available in British Columbia during the 1920s and 1930s, using technology developed during World War I. Aerial photographs may present problems of interpretation depending on the level of vegetation classification desired and scale of the photos, but are useful in recording landscape changes over time and establishing boundaries of land use, and are particularly useful when combined with ground surveys. Stereoscopic aerial photographs taken in 1947 were obtained for the JPRF. An orthophotograph was created from these aerial photographs, and the forest vegetation was classified to current standards using manual and automated aerial photograph interpretation techniques.

5

Appraising the Values of the Forestry

FORESTS PROVIDE MULTIPLE SOCIAL VALUES

The social values of forests differ widely, depending on the culture and social group in question and the roles which forests play in their livelihoods and quality of life. These values are not static, but change over time. Key change factors appear to be the following.

Shortages of, and opportunities to develop, the five main forms of capital (natural, physical, financial, human/individual and social/community capital). Where natural capital such as forest is abundant, it tends to be valued as a resource for conversion to generate other forms of less-abundant capital, and vice versa.

- In a similar vein, the absolute and relative scarcity of particular forest resources, such as timber or biodiversity.
- The availability and price of substitutes for forest goods and services.
- Scientific discovery, education and technology changes.
- Access rights, and resources such as labour that alter capacities to exploit forests.
- An individual's allegiance to specific groups and their shared values.
- Factors that change individual allegiances: campaigning and the political influence of certain groups, tastes and associated communications and media. The prevalence of individual consumerism has changed values away from those of the community.
- Changes in political culture.
- Changes in national development conditions.
- Changes in individual income.

Whilst the table begins with basic livelihood values such as food, and finishes with recreational and aesthetic values, it would be a mistake to assume that there is a universal, rigid hierarchy of needs. For example, for some groups, spiritual values are fundamental, while for others these are much less significant. It is therefore important to identify the relevant values from the point of view of the group in question. Indeed, certain groups define themselves (in part) by the values they hold in common. This is a source of many recent clashes between forest 'stakeholders'. These clashes can be exacerbated when some groups promote groundswells of broader opinion in favour of the values they hold most strongly, as we have witnessed with many environmental campaigning groups or forest peoples' groups. In other words, group values can be influenced by those of others. The outcome is, in large part, a function of power relations, something the forest manager may be able to do little about but must be aware of.

It is also crucial to be aware of the broader trends that affect social values. Policy, institutional and market conditions frequently make the realization of some values difficult. Furthermore, the role of forests in providing such values can be substituted by other sources. The possibilities of substitution tend to be much greater for richer communities than for poorer, forest-based groups. Hence the need for an emphasis on working with the poor.

In developing countries, subsistence domestic products such as fuelwood, poles, fencing and fodder are a priority for forestry, the other priority area usually being food security from forest management or conversion: 'Over 1 billion people (about 20% of the world's population) depend either wholly or to a significant extent on forests, woodland or farm trees for their subsistence needs and/or livelihood'. As Filer and Sekhran (1998) point out, many people who are perceived as rather passive 'indigenous forest dwellers' think of themselves more as forest gardeners and forest developers, with food a number one concern. Forest managers ignore these pressing social needs at their peril.

Some demands for forest values are now far stronger than they were even a few years ago. For example, 'biodiversity' is a word that was coined only in the late 1980s, and the idea of carbon sequestration as a prime value is only beginning to be accepted. Yet now there are all sorts of international environmental agreements, especially the Biodiversity Convention and Climate Change Convention, with significant financial transfer mechanisms emerging to match the political obligations to produce biodiversity and carbon storage values. At present, those who are pressing for such 'global' forest values tend to be the politically more powerful countries and groups such as international environmental NGOs; much remains to be done to ensure that the production of these values also benefits local groups, rather than squeezing out their (more parochial) concerns.

It is quite a challenge for the forest manager to make decisions about the relative weights of the values held by different groups, from local to global, and how to integrate them or how to make choices between them where integration proves impossible.

Three trends are helping here. Firstly, there is the trend for certain markets to recognize some social/environmental values of forest management that were previously unrewarded, for example the market for certified forest products and the emerging market for carbon storage. Secondly, there is development of more inclusive systems of forest policy and management decision-making, such as Round Tables, forest fora, and stakeholder liaison groups, where consultation with different groups about their values is on the increase (even if there is as yet insignificant participation in decision-making and management).

Thirdly, there is the work by economists to categorize values and to develop methodologies to assess them on a single (financial) scale. While these methodologies remain contentious (it is not so much the financial magnitude that counts as who bears the cost and who gains the benefits) and indeed have so far made little difference to major policy decisions, especially those regarding private forests, the economic categorization of multiple values is useful to information science if not yet to decision science.

Broadly speaking, it includes the following.

- Direct-use values: where the value is derived directly from the forest, either in a consumptive manner (timber, nuts, fodder, game, fish, etc.) or a non-consumptive manner (tourism, recreation, etc.).
- Indirect-use values: where the value derives from environmental services such as watershed and social protection, carbon sequestration or biodiversity protection, rather than from the forest directly.
- Passive-use values: where value is accorded by the mere fact that the forest exists (existence value), or for the future possibilities that the forest represents (option value) or for the ability to pass the forest on to future generations (bequest values).

ACHIEVE SECURITY OF FOREST VALUES

People have always planted and managed trees *where* they want, for the goods and services they want, *if* they have the means at their disposal, *i.e.* the technology and resources, and recognized rights, responsibilities, rewards and relationships with other groups.

TRADITIONAL KNOWLEDGE

There is much discussion about the need to reconcile traditional indigenous knowledge with 'scientific' forestry in order to improve local action to secure forest values. It is ironic that, at a time when we are seeking means to sustain multiple values, many traditional management systems that do precisely this are on the decline. For example, Lacandones forest farmers in Mexico cultivate 75 species in a single hectare, and a farmer will need to clear no more than 10 ha of rain forest in his entire farming career. In Sweden, until the nineteenth century, rotational shifting cultivation sustained grain, root crops and livestock as well as timber, and the government saw it as a legitimate form of forest management. As Ghillean Prance has said, traditional forest peoples 'who depend upon so many of the species in the forest will hardly want to destroy more than the absolute minimum necessary for their agriculture. The settler from outside the region, on the other hand, is perplexed by such diversity and tends to destroy rather than use forest diversity'

Traditional forest management systems comprise both knowledge and rules, some of which may be transferable to new SFM approaches. The challenge is partly a democratic one: to admit local groups into policy and management decisions and to ensure that their rights can be exercised. But it is also a challenge to the prevailing systems of knowledge, which by and large are scientific and commercial, to admit such knowledge into the pantheon of accepted practices that guides our forest management.

Rights

Different groups have particular rights associated with obtaining forest values. Some rights directly concern the forest and its use:

- Territorial rights;
- Ownership of trees and other resources, such as minerals, grazing and wildlife;
- Rights of access, use and control (for specific purposes and times);
- Intellectual property rights.

Several problems are commonly associated with rights. One is that different rights overlap on one territory. Furthermore, many rights may not be established in legal documents, especially customary rights.

Other relevant stakeholder rights concern communities and their cohesion:

- Rights to protection of cultural heritage, landscapes and folklore;
- Collective rights, *e.g.* a community's rights to represent itself through its own institutions;
- Rights to religious freedom;
- Rights to self-determination and to development;
- Rights to privacy.

National laws tend to cover specific social issues linked to forests, notably rights to public consultation (planning laws), rights of access (public rights of way) and forest worker rights, etc. Internationally, there is a growing body of relevant legally binding agreements, to which countries may be party, and which eventually find their way into national legislation. These include:

- UN Covenant on Economic, Social and Cultural Rights (1966);
- UN Covenant on Civil and Political Rights (1966);
- UN Declaration on the Human Right to Development (1986);
- International Labour Organization Convention 169 concerning indigenous and tribal peoples in independent countries (1989);
- UN Convention on Biological Diversity (1992). Many new laws are based on recent legal principles deriving from international consensus on sustainable development (such as the 'intergener-ational equity',

'polluter pays' and 'precautionary' principles). Laws on land, environment and forestry are being reviewed in many countries, especially regarding the links or conflicts between formal and customary rights.

Some groups are forceful in exerting their rights and/or have skills and resources to do so. Others are less well equipped. Irrespective of the stakeholders' powers and resources, forest managers are likely to experience problems if they ignore or violate these rights, and indeed claims. An informed approach is better.

This would include anticipating future legal requirements by making provision for best practice in dealing with local groups, and continuous improvement of practice.

REASONS FOR NURTURING SOCIAL VALUES IN COMMERCIAL FOREST MANAGEMENT

Apart from the policy, legal and ethical reasons noted above, there are good commercial reasons to consider social values. Forest managers are increasingly subject to pressures from other interest groups, frequently concerning social values.

- Demands from forest peoples' groups may include greater respect for local populations' rights, carrying out or desisting from specific management practices, and support for their own environmental/social projects.
- Demands from local groups to contribute to social and economic development may include support for local enterprise, employment opportunities, excision of specific areas from management, and use of company infrastructure.
- Demands from trades unions and their initiatives may include greater respect for workers' rights to organize and negotiate, fair wages and benefits, health and safety at work, and the right to skills development throughout the organization.
- Demands to recognize other forest actors' rights to monitor, control and negotiate may include the development of agreements, and procedures to manage conflicts and make compensation.

Unless an active, organized approach is taken to respond to such pressures, forest managers may find themselves facing:

- Slow-downs, strikes, blockades, boycotts, legal battles;
- Damage to forest stock;
- Arson, sabotage or vandalism to equipment and infrastructure;
- Reappropriation of forest lands for cultivation, or migration;
- Development of 'cultures of resistance' among forest-dependent people and their supporters (such as some consumers).

Considerable management skill and time may need to be invested in dealing with disputes, legal challenges and compensation claims, stalled negotiations, 'bad press' and political backlashes and general hostility towards forest managers and companies.

In contrast, participatory approaches that develop people's potential contributions can benefit forest managers, by:

- Improving the reputation of forest managers and companies;
- Uncovering and sharing useful local information;
- Broadening the base of ideas, skills and inputs applied to forestry;
- Efficient apportioning of responsibility, *e.g.* local groups may be better suited to managing recreation and harvesting of non-timber forest products;
- Better understanding of broader social, and thereby market, trends;
- Improving transparency, accountability and therefore trust between all parties;
- Longer-run cost savings and risk reduction.

By taking an active approach to people as well as to trees, forest managers greatly increase the potential social benefits from forest management and their chances of support from others. Such experience should also increase their capacity to anticipate future developments regarding social values, for example in legislation or the market, and thus to gain 'first-mover' advantage from the situation.

If some social values are indicators of market trends (and they frequently are), then it behoves forestry organizations to keep close track of them. Many companies have made good business ventures by diversifying into recreation provision in particular. And it is increasingly clear that there is considerable financial value attached to brand names of certain leading companies which are known to produce social and environmental benefits alongside fibre. For example, the Greenpeace name has been estimated to be worth hundreds of millions of dollars.

CODES OF PRACTICE AND CERTIFICATION STANDARDS ON SOCIAL ISSUES

All initiatives to define SFM principles and criteria cover social values. Some of these are becoming enshrined in legislation, while others face forest enterprises through market relations. Current systems of forest certification specify various social requirements at the level of the forest management unit. Even so, social issues remain perhaps the most contentious of certification standards, and there are differences between standards, particularly regarding the required degree of involvement of different interest groups in forestry and treatment of peoples' rights. There are also differences in interpretation of the standards by certifiers/assessors. Whilst some confine themselves to assessing the local social outcomes of forest management, others assess the roles of local stakeholders in the management of the forest and indeed the enterprise, and call for changes that appear to reflect their own biases. Whilst it is increasingly clear that forestry should not be a principal means of social engineering, and less still should forest certification, it is generally agreed that social standards for forestry will become more stringent in future. Note that some require precise performance thresholds to be met, while others require only that the issue shall be measured. Still others consider that the social issue is a policy concern and require it to be covered in policies only.

BENEFITS OF FORESTS

Trees are necessary for our survival. Through photosynthesis trees produce the gas that we cannot live without: oxygen (O_2). As we breathe in, our bodies take in oxygen and when we breathe out, we release carbon dioxide (CO_2). Trees do the opposite. They take in CO_2 and release O_2. This cleans the air by removing harmful CO_2 so that people and animals can breathe. A tree is often the symbol used to represent the environmental movement. In fact, we adopted the colour green, the plant's colour, as the main adjective to define general environmentalism. Why is the forest such an important element for the environment? There are many reasons why we benefit from forests and there are obviously many consequences that stem from the loss of 13 million hectares of forest every year. Forests stabilize the climate in general. The plants enrich the soil by recycling the nutrients through the shedding of leaves and seeds. They also regulate the water cycle by absorbing and redistributing rainwater quite equally to every species living within its range, which is known as the economy of water. Overall, forests provide perfect habitats for life to flourish on land. They actually contain most of the living species, particularly in the case of tropical forests where up to 90 per cent of the planet's species live. Tropical forests possess the highest level of biodiversity and therefore provide the biggest genes reservoir.

LIVING OFF THE LAND

Earth without forests is a picture that most of humankind presently could not conceive. Forests cover much of the planet's land area. They are extremely important to humans and the natural world. For humans they have many aesthetic, recreational, economic, historical, cultural and religious values. Timber and other products of forests are important economically both locally and as exports. They provide employment for those who harvest the wood or products of the living forest. Herbalists, rubber tappers, hunters and collectors of fungi, nuts, bamboo and berries are able to utilize such resources. Other non-wood forest products come in the form of medicinal compounds, dyes and fabrics. There are many people who are dependent on forestland for their livelihoods. One-third of the world's people depend on wood for fuel as a significant energy source. Surveys in Cameroon, Cote d'Ivoire, Ghana and Liberia found that forest wildlife accounted for 70 to 90 per cent of the total animal protein consumed. Some indigenous peoples are completely dependent on forests. As well as providing a home for some people, the forest environment provides a popular setting for eco-tourism, which includes hiking, camping, bird watching and other outdoor adventure or nature study activities.

Protection from Natural Disasters

Trees protect the soil against erosion, and reduce the risk for landslides and avalanches. They may increase the rate that rainwater recharges groundwater, as well as control the rate that water is released in watersheds. They help to sustain freshwater supplies and therefore are an important factor in the availability of water, one of life's basic needs. When rain falls, some may sink to the ground, some may run off the surface of the land, and flow towards the rivers and some may evaporate. Running water is a major cause of soil erosion.

During heavy rains, flooding may occur, filling the waterways with eroded soil. The silt clogs these waterways, cutting off water sources for plants and animals during the dry season. Silt may also fill reservoirs created by dams, reducing its ability and future capacity to generate hydroelectricity and provide irrigation. The removal of forests causes nutrient loss in the soil especially if the period between harvest isn't long enough. Scientists have recognized that trees can also serve as a tool in the reduction of storm water runoff. The incorporation of trees and other vegetation costs five to ten times less than using solely manmade storm water infrastructures. The leaves on trees keep large quantities of rain and snow from falling to the ground and tree roots absorb excess surface water, thereby stabilizing ground soil. Street trees provide the greatest annual benefit in avoiding storm water runoff by diverting 327 gallons of water compared with the 104 gallons averted by park trees.

Helping the Climate

Researchers have found that trees help the urban ecosystem by decreasing air temperatures. Studies indicate that a 10 percent increase in tree canopy cover results in a one to two degree Fahrenheit reduction in air temperature. In addition a one-degree decrease in temperature will reduce the possibility of smog by 6 percent. Furthermore, increased tree canopy coverage protects urban dwellers from harmful effects of ultraviolet (UV) radiation. Strategic planting of trees can also increase a city's energy efficiency. Research conducted since the mid 1980s has quantified the energy saving potential of urban forests. According to the Energy Information Administration, household heating and cooling cost consumers 180 billion US dollars in 1987. Studies have found that a 25 feet tall tree could save 10 to 25 US dollars annually on these energy costs alone. Because trees release cool vapour into the air during photosynthesis, the need for artificial cooling devices is reduced. In fact, according to one study, the air-conditioning savings from a deciduous tree near a well insulated home ranged from 10 to 15 percent, while an 8 to 10 percent savings was reported during peak cooling periods. Landscape vegetation around individual buildings can also result in heat savings of 5 to 15 percent savings, and cooling savings of 10 to 50 percent.

It has been observed that one of the largest energy fluxes at the earth's surface is that due to evaporation by trees. Heat is absorbed by trees for transpiration of fluid, and later released into the upper atmosphere. The fluid involved is groundwater flowing up the trunk of the tree. Increased annual runoffs from deforested areas in the Amazon support this attribution.

Purification of the Air

Forests affect the climate and are an important source of oxygen (O_2), although they play a lesser role than once thought.. Rain forests serve as an important filter for carbon dioxide (CO_2), a greenhouse gas that contributes to global warming. The Amazon region alone stores at least 75 billion tons of carbon (C) in its trees. When stripped of its trees, rainforest land soon become useless and inhospitable because the soil lacks the nutrients to support any kind of agriculture.

Regeneration of a tropical rainforest may not be possible or, when it can occur, it may take hundreds of years. Research continually reveals that trees benefit urban communities in a number of ways. First with respect to air quality, trees remove damaging pollutants from the atmosphere, and replenish it with O_2. Through the process of transpiration and photosynthesis, trees sequester grams of ozone (O_3), sulphur dioxide (SO_2), nitrogen dioxide (NO_2), and carbon monoxide (CO) every hour, amassing several tons of carbon storage each year. This carbon sequestration process in turn reduces the harmful effects of these noxious gases that cause global warming as well as lung-related ailments. Researchers have also been able to quantify the value of this carbon removal through the use of a carbon storage and sequestration model called UFORE-C. In fact utilizing the figures economists employ to estimate the effect pollutants cost society, one research ecologist was able to compute carbon sequestration into a tangible "dollar-value.".

CAUSES OF DEFORESTATION

There are many causes of deforestation however harmless they may seem. So much damage can be done by even a single chainsaw, because behind those chainsaws are huge companies that care only about demand and profit, and forests are needed to supply this.

Commercial Exploitation

The first and most important cause of deforestation is wood extraction. Wood has always been a primary forest product for human populations and industrial interests. Since wood is an important structural component of any forest, its removal has immediate implications on forest health. Intensive harvests can lead to severe degradation, even beyond a forest's capacity to recover. When the soil has been stripped of its nutrients, farmers move further into the forests in search of new land. Shifting cultivation is one of the most unproductive uses of farmland, and a major cause of degraded land where forests cannot regrow. In eastern India, this agricultural practice called "jhumming," has laid barren previously fertile tracts of the hillside. Timber is one of our most precious as well as used resource. We use it to build our houses, furniture and stock our fireplaces. This heavy demand fuels the destruction of our forests at an unsustainable rate. For every tree that is logged, 27 are killed or damaged in the process. In small doses logging isn't too bad, and over 30 or so years the forest will grow back. But with today's demand for wood the areas that are logged are too large, and this causes permanent destruction of the forests. The impact of the timber trade is generally greater than has been claimed in the past. The North plays a key role in many of the factors leading to forest decline.

Commercial forestry is the leading cause of deforestation in the world's temperate regions. The forces of large global markets for wood and wood products drive the scale of logging activities such as clearcutting. The source of demand is increased consumption by North America and Europe, not population growth. Again, transportation routes have a role, opening up new areas for natural resource exploitation. Privatization of natural resource industries has led to decreased regulations regarding timber harvesting. Multinational corporations dominate trade in wood. Most of these companies were organized in the US. Multinational companies for whom improvement of forest practices is not a priority often export the timber in an unprocessed state out of the country of origin.

Mining for precious resources also plays a major role. There are many forests that hold fair amounts of Earth's resources, such as iron ore, copper, oil and other precious metals. Many mining methods such as strip-mining and strong-force hoses break down the earth and cause major erosion. The mining sites are large and many trees need to be demolished to make way for them. When nothing is left to be mined, there is little chance of the forest growing back because of erosion and the lack of nutrients in the soil, which was churned out during mining.

Development of Degrade Forests

People destroy or degrade forests because, for them, the benefits seem to outweigh the costs. Underlying causes include such issues as poverty, unequal land ownership, women's status, education and to some extent, population. Immediate causes are often concerned with a search for land and resources, including both commercial timber and fuelwood. In many areas, rural households rely solely on fuelwood collected from the forest for their domestic energy supply. The roads that are built into the rainforest encourage and provide access for settling activities. In north-east India road building is often wrecking havoc on the forests. A road is cut through a hill face and the first loss is of the trees along its trace. The debris is thrown down, destroying the trees below, leaving a trail of dead or wilting trees in its wake. This debris enters the valleys, pushing up the levels of streams and rivers, causing siltation and floods.

There are many government agencies with policies that are uncoordinated in nature. Long range planning is not undertaken; and the Amazon is greatly affected by forces outside of the region. Some of the causes of migration to the tropical forests are population growth and political persecution. The settlers clearing and cultivating the land do not have the knowledge and experience of indigenous peoples of the forests and are unable to utilize the land effectively or sustainable. The process of shifting cultivation is accelerated and as a result the forest doesn't have enough time to recover. Tropical rainforests are truly under the assault by humans.

Cattle Ranching

Perhaps the worst culprit of deforestation, at least in the Brazilian Amazon, is cattle ranching, accounting for 38 percent of deforestation in this region. Cattle ranching involves hundreds and thousands of cattle grazing on expansive areas in and near forests. Because the forest soil isn't adapted to these conditions, it isn't long before the area becomes unproductive. So cattle ranchers expand their grazing area, leading to more destruction. After the grazed land is left, the forest is very unlikely to grow back due to the stripping of the soil. Cattle ranching each year in this manner destroys an estimated 5700 square miles of rainforest alone. The Brazilian government subsidizes some of the cattle ranches that exist on converted forestland. The land is unproductive. Much of the demand for the beef comes from the fast food hamburger market, which is more concerned with quantity, than quality farm raised meat.

Agriculture, Ecosystem and Forests

Monocultural forestry simplifies the ecosystem, leaving it vulnerable to disease and other environmental factors. In the tropical forests of the world, the clearing of land for agriculture and livestock are the primary activities resulting in deforestation. The main cause is unequal distribution of land. 4.5 percent of Brazil's landowners own 81 percent of the country's farmland, and 70 percent of the rural households are landless. It seems that these conditions cause people to encroach on, penetrate and modify the forests. Governments have an important role in these processes. In many countries in Asia and Africa, where family farms are still prevalent, the breakdown of large joint-families is causing uneconomic divisions to existing farms. These inefficiencies, in turn, put pressure on farmers to sell their land for development, and it turns whole farming comminutes into new developments.

Poverty and Inequality

Another cause of the ecological crisis of the present is social inequality. Gender inequality is one of the more powerful forces at work, which exists in virtually all of human cultures. The natural world is often portrayed as feminine, in terms such as "Mother Nature," "virgin forest," "exploitation" and "rape of the valley" that are used to describe elements and uses of nature and serve to perpetuate this harmful attitude. Human society's attitude with regard to the status of women makes an important contribution to environmental degradation and deforestation. Although it is easy to assume a strong connection between population growth and deforestation, some research indicates that the problem is more complex. It involves non-demographic mechanisms resulting from credit and capital market failures, lack of suitable mediating institutions securing property rights, wretched poverty, uneven land distribution, consumption patterns in developed countries, greedy multinational companies, ignorance and bad management by colonist or frontier land, and so forth.

One of the underlying causes of human exploitation and consumption of forests and other natural resources is human tradition and beliefs. One source of such belief is Christianity, whose dominance in the Americas and Europe has important consequences for natural resources. Christian's attitudes are of anthropocentrism. The dominant power on the planet is humankind. The first human, Adam, gave the animals their names, shows this kind of dominance. Modern science and technological changes began in the name of Christianity. These beliefs created the attitudes, traditions and activities that enable us to be responsible for the destruction of nature that is occurring in the forests and the rest of the world. It is then possible that if certain Christian beliefs were different, human attitude towards nature would be that of conservation not exploitation. Such contrasting beliefs could include a god or gods that exist on Earth or even in trees, or that humans are reincarnated as plants and animals.

RIGHTS OVER FOREST RESOURCES

The tribal people are facing serious problems with regard to utilisation and rights over forest resources. Due to the increasing pressure on forests by various interest groups, there is a corresponding pressure on the tribals to reduce their dependency on forests. This is creating serious situations of conflict, as tribal life is symbiotic with land and forests and their livelihood and culture are based on their relationship with the natural wealth around them. The tribals are being harassed for using forestlands and being evicted in many places. Such reports have come from places like Khammam, Visakhapatnam, Vizianagaram, Adilabad, Srisailam and other places. In Khammam in one particular village, the forest and police departments allegedly branded the tribals on their shoulders as an indication that they were destroying the forests. There are ambiguities in forest-revenue land demarcation. In some places like Nellore district, a lot of land on Velugonda hills is indicated as 'poramboku' in revenue records and as RF as per the forest department. These lands do not have any forest growth. The FD is taking up palm oil plantation in these RF lands (which is not a forest species). However, landless tribals are booked in criminal cases or prohibited from using these lands for agricultural cultivation. These lands should be given to tribals with pattas for cultivation.

In some areas like in Visakhapatnam district, the lack of clear forest boundaries is making tribals vulnerable to the exploitation of both the forest and revenue departments. A joint survey and demarcation of boundaries by both departments should be immediately taken up to arrest these conflicts. Such Joint surveys need to involve the villagers at various levels. The Forest-Revenue Boundary dispute is a perpetual problem in Adilabad and Warangal districts, leading to booking of cases by the Forest department and tension in these tribal villages. In Visakhapatnam Agency and Nallamala areas, there is the unique problem of Enclosure Villages. There were many tribal villages that were not enumerated in the forest surveys. Due to such sheer negligence, the villages were not given revenue status and to this day, they do not have pattas for their lands. They face constant harassment from local forest officials, as they do not possess land records. Recognizing these enclosure villages and issuing pattas to tribals should immediately resolve this problem. In Buttapur (Adilabad district) and in Nellore Dt. (Yanadis), the tribals were given lands decades ago under the social forestry scheme and are cultivating there. But due to lack of pattas, they are being harassed by the police and forest departments and also do not have access to bank loans as they cannot prove their ownership. These tribals have to be given pattas as promised.

Attacks on tribals, their properties and livestock by wildlife are not compensated by the Forest Department. Several cases are pending where tribals have been either killed or disabled and yet have not received any monetary compensation as due to them under the Wildlife Act. The tribals should not be prohibited from entering the forests to collect NTFP for their domestic requirements, like firewood, medicinal herbs, food, or agricultural and housing material.

The tribals should not be defined as 'encroachers', as is being projected by the forest department and also as indicated in the Circular of IG Forests, MoEF dated 3rd May 2002. In districts like Visakhapatnam, East Godavari and Vizianagaram, where there is high prevalence of podu cultivation, the tribals are facing threats of eviction from the forest department. There was a notification issued by the government of A.P in 1987 ordering that pre-1980 settlements will not be evicted until further orders. This should be implemented. The JFM programme now renamed as the Community Forest Management programme of the A.P Forest Department has caused grievous violations with regard to tribal rights. One major violation is the displacement of tribals from their podu lands by reclaiming them back into the forests through the JFM programme. The official reports of the forest department and the World Bank (which has funded the project) reveal that 37,000 hectares of forestland has been reclaimed back from the people§. The APFD further states its intention of displacing tribals in G.O. where it calls for voluntary surrender of lands by involving NGOs to motivate people under the CFM programme. The APFD has come up with a proposal for rehabilitation of tribals displaced from their podu lands through a monetary and schematic approach. This should be condemned as it bypasses the issue of eviction and rights of people over their podu lands. Their act of reclaiming lands under JFM programme and the APFD's forestry project under World Bank assistance has to be scrutinized closely by the Commission.

Samata and a few other NGOs raised this issue by writing to the World Bank on its violations of its own Operational Directives 4.2 and 4.3 in its financial support to the APFD. This has led to the Bank's insistence on an R&R policy to be proposed and implemented by the APFD in its CFM programme In Srisailam area, the Rajiv Gandhi Tiger Sanctuary has led to eviction of tribals from their original homes. They have not been properly settled so far. The concept of EDC (Eco Development Committees) that was introduced is working to the detriment of the tribals, as the objective of this programme was to reduce the forest dependency of the tribals. The tribals are given income-generating programmes and alternate sources of firewood by the forest department. This is an artificial mechanism that is not sustainable in the long term.

In the Srisailam Tiger Sanctuary area, the Chenchus, who are traditional hunter-gatherers, go into the forest everyday for all their needs. They are being harassed by the forest department for trespassing into the sanctuary. The groups who came from the chenchu area felt that the Chenchus should atleast be given identity cards to prevent harassment from forest department and the police department who mistake them for naxalites. Another serious problem with regard to Sanctuaries is the settlement of people's rights. The Settlement Officer is the Conservator of Forests whereas the forest department is one of the interested parties in the land acquisition for the sanctuary and hence it is not appropriate to make the forest officer the settlement officer**. It should be the revenue department (District Collector) who should be authorised to settle the people's rights. G.O.No.112 which calls for involvement of private industries like ITC, Reliance and others for taking up commercial plantations through the VSS should be withdrawn immediately, as this is a backdoor method of allowing private industries to enter forest and tribal lands. Such tripartite agreements between the government, the industries and the tribals can never provide a level playing ground for the tribal people and make them more vulnerable to the exploitation of private industries.

The historical injustices to the tribal people in this region and that of the tribals in southern Orissa due to construction of several 'development' projects and industries like Nalco, HAL, Sileru, Machkund and others which have displaced tribals in large numbers without any rehabilitation, and has forced them to migrate in search of livelihood. They are now being treated as encroachers and criminals in forestlands. This definition should be condemned and Commission should recommend proper mechanisms for resolution of conflicts.

The Commission should strongly recommend the orders passed by the Ministry of Environment and Forests in 1990, on the recommendations of the Commissioner of Scheduled Castes and Scheduled Tribes, to ensure that such conflicts are resolved. Infact, the first two orders alone show a way forward to settle disputes concerning forestlands. All the groups consulted strongly felt that the pressure on forests and the destruction of forests is taking place more due to increase in industries and private commercial activities like Mining, paper-mills, timber smuggling and the growing non-tribal population settling down in Agency areas, than due to the podu cultivation practised by tribals. Eco-Tourism is also causing destruction of forests. For example, the A.P Tourism Department built huge infrastructure close to Farhanbad, near Mannanur (Srisailam) for Tourism purposes in the middle of the protected area and in Araku, Borra and Anantagiri (Visakhapatnam). While tribals are prohibited from entering these forests for their survival needs, it is totally unjust to allow tourism, which is causing a lot of degradation. Hence, the constitutional provisions of the Fifth Schedule, the Samatha Judgement and the PESA Act should be strictly implemented following the right spirit with regard to industries and non-tribal settlers, which includes the Tourism Department.

The Tourism Department also conducts tours through tribal villages, especially in Chenchu settlements, and exhibits them like museum pieces. The groups felt that this degrading exercise further underscored the lack of empathy among the various government departments towards the tribals. Because of a shift in the policy of the government, profits to G.C.C. are more important than benefits to tribals. Although there are 32 items in the procurement list of GCC Ltd, it procures a limited number of commodities and at very low prices compared to the market rates. The tribals face several constraints in collection of NTFP and marketing these items because of the monopoly rights enjoyed by GCC Ltd. They do not have the right to sell their produce to private traders even when their rates are higher. Hence, a long-standing demand has been for removal of the monopoly restriction of GCC Ltd. and instead a minimum support price should be provided by GCC to protect the tribals from exploitation of traders. The monopoly of forest department over Beedi leaf collection and sale should be removed. The tribals or VSS should be given 100 per cent rights over income from beedi leaf and also the right of sale, with the forest department providing the support price.

INDIAN FOREST LAW AND POLICY

After Independence, however, and according to the newly enacted Indian Constitution, forests were placed under the so-called 'state list' according to which state legislatures have a primary right to make laws. In 1976, the Indian Forest Act was added to the concurrent list of the Constitution of India, giving the Central Government and states shared responsibility and control over forest matters. Thereby the Government of India does have the power to legislate on forestry issues but only after consulting the states. The responsibility for administering the forests lies primarily with the state government. The relative position of the Government of India depends on several factors, such as its contribution to the state budget and state polities. The balance of power between central and state governments has remained a key issue in forest management ever since. After 1947, in a post-independent India, commercial exploitation and degradation of India's forests increased dramatically. Indeed, the 1952 National Forest Policy set out guidelines which were, for the most part, directed towards the supply of cheap timber and non-timber forest products for state-sponsored industrialization and modernization. In 1980, with the passage of the Forest Conservation Act, the central government reasserted some of its control over forest-based resources. The Act restricts the state government's power to de-reserve a forest, and it restricts the use of forestland for non-forestry purposes without the prior approval of the central government. Unilateral decisions from the centre have prevailed from then onwards. Only six centres were set up to monitor forests and conservation (as per the Act), which is obviously insufficient for effective regional implementation in a country the size of India. Thus the Forest Conservation Act of 1980 has been problematic for a number of reasons, and has achieved little improvement in the 'conservation' of India's forests.

The National Forest Policy of 1988 envisaged people's involvement in the development and protection of forests *for the first time*. It stipulated that the requirements of people living in and near forests for fuel wood, fodder and small timber should be treated as the top priority, and forest communities should be motivated to identify themselves with the development and protection of forests from which they derive benefits. A primary task of all agencies responsible for forest management, including the forest development corporations, should be to engage tribal peoples closely in the protection, regeneration and development of forests, as well as to provide gainful employment to people living in and around the forests. However the 1988 National Forest Policy has never been translated into law. It remains essentially a broad statement of government intent and does little in the way of specifying any legal rights or duties owed to forest communities-especially the tribal/indigenous peoples.

NATIONAL FOREST POLICY OF JOINT FOREST MANAGEMENT

One outcome of the National Forest Policy was the launching of Joint Forest Management (JFM) initiatives by all the States and Union Territories. However, there have been no accompanying changes in the legal framework for JFM, and in almost all states it has been implemented through administrative orders, which do not take the existing infrastructure into consideration. Forests continue to be looked upon as the property of the Government and the delegation of management responsibilities to local communities is merely an administrative arrangement, falling short of any real devolution of power to local communities. As an illustration. Prakash Kashwan has looked into JFM from various points of view and highlights some other problems. One of the most important findings is that JFM has been allowed only in degraded forest, but in order to bring more communities under the ambit of such programme, communities have had to actively pursue degradation of natural forest. Furthermore, JFM requires the demarcation of boundaries between villages, often for the first time, but poor implementation has resulted in many boundary conflicts. Joint Forest Management (JFM) in the state of Jharkand.

- JFM in the state of Jharkand is based on the premise that people are to blame for the destruction of the forest. The objective, therefore, is gradually to diminish people's dependence on forests.
- The Resolution to introduce JFM has been made unilaterally by the Forest Department. There is no scope for consultation with the people, no effort at consensus building, and no mechanism for dissemination of information.
- The resources required for villagers' needs are to be extracted from the protected forest according to the micro/annual-working plan. There is no mention of the existing rights of the villagers to use the produce from these forests. This is a gross violation of customary rights.
- A 33 per cent share in net profits was promised. Now the promise is to share 90 per cent of net profits with the forest dwellers. The profits however, are from commercial productions, which are not based on local needs, thus indicating the focus on commercialization of forest. The JFM Resolution in Jharkhand is simply an extension of the undivided Bihar Forest Policy (Jharkhand used to be part of Bihar before it became a separate state), which adopted Joint Forest Management in 1990. Jharkhand Forest Department adopted the JFM Resolution on 27th September 2001

- The Forest Department claims to have formed 2034 Village Forest Management and Conservation Committees and Eco-Development Committees managing an area of 5,589 sq. km. in a remarkably short time span. However, local people emphatically rejected JFM. Apparently, the figure is manipulated and incorrect.

Eviction of communities as Forest Encroachers

Indigenous peoples feel that measures that were put in place in the name of protecting the forest have only deprived them of their resources. The JFM initiatives have failed from the point of view of indigenous peoples, and there has also been widespread eviction of communities from forests. The forest department uses any means available to alienate the forest communities from the forest, reduce their dependence on the forest in the name of forest protection and conservation, and invite private partners and multinational companies to negotiate directly with village committees to access forests for carbon trading, Clean Development Mechanisms and trade in eco-services.

The direction of forest management is thus very far from anything that takes traditional forest related knowledge into account or that recognises the rights of indigenous peoples to maintain their traditional forest practices; rather, it aims to separate them from the forests, and sets up an externally imposed process of commodification and marketing of forest goods.

Indigenous Peoples' Rights and Forest Management

The Government of India is not a signatory to the ILO Convention 169 and as such does not recognize indigenous peoples. It does recognise "Scheduled Tribes", but not all indigenous groups are scheduled. Some of the indigenous groups are recognized by the Government of India as Scheduled Tribes and enjoy a degree of special status. Some of the areas where the Scheduled Tribes live have been demarcated as Scheduled areas and under the Indian Constitution Sixth Scheduled provisions, Autonomous District Councils have been set up for local administration of the Scheduled Tribes. In theory there are some measures to protect the interests of Scheduled Tribes. The Constitution empowers the President of India and state governors to withhold any law considered detrimental to tribal/indigenous peoples' interests in Scheduled Areas. Items V and VI of the Indian Constitution give special privileges to the Scheduled Tribes and the Panchayats Act (Extension to the Scheduled Areas), 1996 (PESA) is also designed to offer some protection. However, in practice most of the laws restrict the rights of and control of forest communities. Specifically, the Indian Forest Act (IFA), Forest Conservation Act (FCA) and Wildlife Protection Act (WLPA) continue to be used to hound forest dwellers.

JFM and CFM are present in most Indian states, but are simply plantation programmes related to timber and other economic products, which have nothing in essence to do with forests. They should include Village Forest Committees directly representing the community living in or using the forest. These programmes are operational to varying degrees in different places with varying degrees of participation of indigenous peoples depending upon the local situation. The involvement of indigenous peoples is focused on providing them with a share in the economic benefits and *de facto* also acts to wean them away from their dependence on forests. In other words, integral to these programmes is a subtle (in some cases blatant) redefinition of indigenous peoples' relationship to the forests. The degree of incorporation of TFRK is varied, if there is any at all; the expertise is provided by the forest department, sometimes in partnership with NGOs.

Indigenous peoples' rights to the forests are progressively minimized. In some cases, indigenous peoples do receive some economic benefits (though in others non-indigenous peoples dominate the Village Protection Committees), and in the context of the takeover of resources necessary for basic survival and consequent deprivation, these economic benefits at times are seen as a steady source of survival income. This in certain cases has superficially reduced tensions and conflicts between the forest department and the indigenous peoples as a sense of co-existence is manufactured. However, this also means that the forest department is able to stifle protests and resistance. In many places, where the indigenous peoples' political consciousness is well developed, they keep away from or oppose these programmes leading to conflicts, as has happened in parts of Madhya Pradesh. India's report to UNFF3 (2002) describes measures taken to make use of indigenous capacity and local knowledge in participatory forest management. Clearly, the picture painted by this report is not matched by the situation on the ground.

FOREST MONITORING

Forest monitoring is very useful to detect changes due to climate change, natural disturbances or human activities. It has become a requisite in the context of climate change mitigation in particular, in relation to deforestation and forest degradation. Due to the potential benefits that accurate carbon monitoring may bring within the REDD+ framework, monitoring has developed greatly over the past few years and requisites on accuracy and acceptability have increased. For monitoring in general to be successful, it needs to have clear objectives, be as simple as possible, and benefit the people that invest time and/or money in it. However, many times the objectives may be clear, but the activities that are needed to meet those objectives may be vague. This may be due to lack of experience or lack of certainty on how climate will change and how this possible change will affect different components of the forest and forest management. A tendency exists to want to monitor everything that might possibly change, resulting in impractical and expensive monitoring proposals.

As a result, monitoring for the impacts of climate change on the forests and people related to the forest is still just emerging. Monitoring helps us to identify changes and evaluate tendencies. Monitoring does not necessarily tell us the reason for these changes and tendencies, unless previous research has established such causal links. The next step would therefore be to analyse whether such changes correspond to changes in climate characteristics and then, to analyse whether such tendencies are negative or positive for the forest and the forest managers and whether actions can be taken to reduce the negative consequences and increase the positive ones.

Current discussions on the implementation of REDD+ are occurring at the national level, however most of the monitoring experience has been obtained at the forest management unit level. While monitoring needs at these levels differ, they are highly complementary and any carbon monitoring system should consider linking these levels. It is very important to include all stakeholders to ensure agreement on the methodology and the variables to be monitored. The involvement of local actors has been shown to have two advantages; it is cheaper and creates greater ownership of the monitoring results. In spite of the importance of monitoring for SFM and for preparation of responses to climate change, it still is not a common practice. Particularly in developing countries, few forest managers have the resources (human and financial) to implement these assessments.

MONITORING OF CHANGES

Adaptation of forests requires in the first instance the identification of the changes that may occur and to which adaptation may be necessary or desirable. Although in general terms forest change scenarios can be developed based on global and regional climate change projections, the exact changes that will occur are not well known. There are several reasons for this uncertainty; the uncertainty in the climate change models in general, the scale at which climate change projections are made, the 12 inherent adaptive capacity of species and the communities they are in and the effect that interactions between species may have on adaptive capacity. In some areas, the changes that have been projected are drastic. The northeastern Amazon, for example, may lose most of its forest cover because of massive forest dieback due to droughts, giving rise to savannah vegetation. However, the rate of change and the exact result is not that clear. Other areas may follow suit at different rates and with different results. It will be difficult for forest managers to react to these changes, especially if it is not clear when and how these changes occur.

Adaptation strategies will need to include monitoring systems on climate, vegetation, fauna and essential non-biological components of the forests such as water availability. Without such monitoring systems, the forest manager will be grappling in the dark when making management decisions. In forestry, such monitoring systems are important, particularly because of the long time lapse between management actions and forest response. For this reason, permanent sample plots (PSP) are an integral part of SFM. Their main contribution to SFM has been a better understanding of the dynamics of forests and plantations. PSPs have been used for stock-taking (both at a national scale and in continuous forest inventories), for monitoring of changes in managed and unmanaged forests (*e.g.* in certified forests to monitor changes in species composition and structure), and for research purposes (*e.g.* the effect of silvicultural treatments and harvesting on species composition, structure and biodiversity). PSPs are less useful to measure changes in the diversity of fauna, impacts of forest operations such as harvesting and impacts on ecosystem services that go beyond the forest plot boundaries (for example water flow).

In countries that have long-standing experience with PSPs and a good network of meteorological stations, PSPs may provide a good contribution to the analysis of the effects of climate change on forests. While PSP are good instruments to detect changes at the stand level, forests are also influenced by changes that occur on a landscape level, *e.g.* water quality affected by sedimentation. To detect such changes, a combination of remote sensing techniques and a network of PSPs is probably the most appropriate strategy: remote sensing to detect changes in forest areas, and PSPs to detect changes in forest quality. Since remote sensing images and their interpretation for forest management is relatively costly for the forest manager, such monitoring is best carried out by organizations or associations that are responsible for larger areas or a group of stakeholders. This will require, that all potential users of the monitoring information agree on a common set of variables that are useful for forest management decisions and should therefore be monitored. An important part of monitoring systems is the database and processing of the data. This usually requires major investments in human resources but some companies have been able to develop their own computer hard and software that allows for quick data storage and analysis.

Monitoring Techniques of Animals

The techniques used monitor animal populations, particularly the larger mammals, depend on the objective of sampling. More local research is necessary to identify the techniques and variables to be sampled in specific cases (*e.g.* for a particular species in a defined region). Climate change can shorten the life cycle of insects, increasing their reproduction rate and the risk of infection and damage. Traps in sampling points can help to detect rapid increases in population sizes. The traps will need to be specific for the insects to be monitored, have appropriate bait and be placed at the right position in the forest. Turchin and Odendaal (1996), found that one funnel trap, used to trap southern pine beetles in the united States, were good for covering an approximate area of 0.1 ha.

Some insects are more ground related (*e.g.* dung beetles) while others (*e.g.* butterfly families) may fly in open or closed forest areas. Although these latter insect groups have not been related to pests, they have been successfully used to identify changes in forest structure and composition related to fragmentation and tree harvesting, and may be useful to detect forest changes due to climate change. Specific dung beetles may be related to specific mammals and butterflies have been related to openness of the forest and may be an indication of dieback. Further research is needed to fully understand these relationships. Larger animals may be trapped (as in the case of small rodents), or counted visually using walking transects. Animal tracks may also be used as an indication of the presence and abundance of species. However, care should be taken that sampling density is sufficient to formulate robust conclusions. Steel *et al.*, (1984) concluded that three repetitions of a 2 km transect was sufficient to determine species abundance, richness and diversity of large animals, but it was more difficult to estimate small mammal richness and diversity. In general, design of a monitoring system requires expert knowledge, but local communities can be trained as parataxonomists to implement the monitoring.

Due to the need for additional information on species behaviour and preferences, as well as the relatively high time investments needed for animal monitoring, it is important to identify, as early as possible, those animals (and plant species alike) that are more susceptible to climate variations. Abundant animals may be easier to monitor, but many of them may also be less susceptible to changes in climate and the environment. Usually the species with a small range and short generation time are more responsive.

Forest Fire Monitoring

Monitoring of forest fires contributes to our knowledge on the extent of deforestation and forest degradation. Such monitoring is traditionally done through patrolling forest areas and operating watchtowers. The development of remote sensing techniques has made it possible to come to ever more accurate and timely information on forest fires, above all in large uninhabited areas. Laneve *et al.*, (2006) estimate that if images can be obtained at a sufficient spatial resolution to detect 1500 m^2 fires at 30 minute time intervals, this will be sufficient to reduce the number of large fires in the Mediterranean forest of Italy.

For small forest land holders and many communities, such technology is not available and even the construction of towers may be too high an investment. Patrolling, however, has shown to be an effective way of forest fire prevention in community forests in Guatemala. During the last decade several proposals have been made to set up fire detection systems using wireless sensors, but most of these have not emerged from the experimental phase, possibly due to costs and the problem of maintaining the network.

CAPACITY OF FORESTS TO RESPOND TO CLIMATE CHANGE

The adaptive capacity of forests, for the purpose of this document, is understood to be the inherent ability of the forest to adjust to changing conditions, moderating harms and taking advantage of opportunities (Locatelli *et al.*, 2010). Thus, strengthening of the adaptive capacity is oriented at increasing the resistance or resilience to changes but may also include adapting the forest to new conditions by facilitating changes in the system (*e.g.* by species introduction). In general, strengthening the adaptive capacity of forests aims to maintain, restore or enhance forest area, biodiversity and forest health and vitality. Many of the actions oriented towards mitigation of climate change through REDD+ have a strong potential for synergies with actions oriented at strengthening the adaptive capacity of forests, in particular if such actions consider ecological safeguards, such as biodiversity conservation.

Experiences in strengthening the adaptive capacity of forests to climate change have been more widespread in plantations and agroforestry systems. These systems tend to have a simpler structure and composition that makes it easier to detect changes due to climate change and to design and implement adaptation-strengthening mechanisms. This is much more difficult in complex natural forests, in particular in the tropics. However, because of their simplicity, these systems may also be more vulnerable and therefore the need to look at adaptation options is greater. Interestingly, several of these adaptation activities are oriented towards making these (agro) ecosystems more diverse.

Maintaining Forest Area

Larger forests usually have greater species diversity and cover a greater variety of sites, thus reducing the risk of losing the whole system if climate change negatively affects several species or specific site conditions. Forest management appears to be the solution; both wellmanaged protected areas and well-managed community and private forest concessions in Guatemala and the South of Mexico have shown to be more effective in avoiding deforestation, fires and forest degradation than areas poorly managed areas. This probably goes beyond purely economic considerations. Land and forest tenure, recognizing the benefit of maintaining forests and joining forces with other forest managers are all important requisites. Size of forest area also seems to be important, for both individual landowners and community or multiple owners. Ecologically, larger forest areas show less edge effects, while from the management point of view, larger sizes allow for economies of scale. Managing natural forests often is recognized as a claim on that forest. If good relations are held with local people, such claims are well respected.

Owning forest but not managing it, has often resulted in unauthorized entry by third parties for the extraction of timber and NWFPs, or for conversion to agricultural land. Managing the forest but not entertaining good relations with the neighbours has often resulted in forest use conflicts, at times ending up in armed conflicts or burning of parts of the forest estate. Such relations are more important in large forest tracks, since in these it is harder to establish continuous human occupancy. Good forest management normally includes fire, pest and disease management.

Of these, fire management may be the most significant in maintaining the forest area, although serious pests, such as the mountain pine beetle in pine forests in North America, may also contribute to substantial forest loss. Managing the forest may be costly and income from the sale of one or more of its products may not off-set the extra cost of management. However, often, costbenefit analyses compare conventional operations (without much strategic planning or considerations for biodiversity or forest dependent communities) with managed operations.

From a private forest owner's point of view, this may be reasonable, but in practice, this approach has been used to justify continued conventional harvesting operations, giving the forest sector a poor image and increasing the pressure on governments to impose stricter regulations. In countries with greater willingness of the private sector to participate in 15 improving forest management, a series of alternatives were found to either make forest management attractive or propose other forest-based income solutions. Usually this was done by a carrot and stick approach: if a forest manager does better than the legislation requires, they receive subsidies, discounts on taxes and are a preferred provider of specific ecosystem services. While these approaches in theory seem to be very promising, in practice they have not had the expected outcomes. This is partially because of the high financial and administrative cost to actually obtain the carrots and because forest owners and managers were not aware of the existing opportunities.

Management of forests, therefore, should go beyond the mere planning of protective or productive activities and not depend on one single form of financial income. Forest managers need to be informed and aware of local, national and international opportunities for income generation. New opportunities may be through tax or fee discounts for good forest management (as the case of harvesting fee discounts for certified forests in Peru), payment for environmental services (*e.g.* Costa Rica and Mexico) or niche markets (*e.g.* markets for specific NWFPs and carbon).

Due to the relatively limited demand of these above mentioned products and services and with only a few exceptions (*e.g.* Brazil nut gathering in Bolivia, Stoian, 2004), none of these have been able to single-handedly pay for management and protection of large natural forest tracts. Only where combinations of products and services were obtained has management of natural forests become a serious land use competitor. New opportunities may also work for forest plantations, where payment for environmental services, such as carbon sequestration, may at least partially off-set the initial establishment costs. However, selling sequestered carbon at the end of the rotation has little effect on the overall profitability of plantations, due to the low carbon price and high interest rates. The present value of these future sales is very low and rarely will change a non-profitable exercise into a profitable one. Selling less carbon earlier on during the rotation, or shortening rotation length, are two options that may make timber plantations more attractive. Establishing tree plantations not for timber, but only for carbon or other environmental services, as yet has to show its profitability for the forest owner or manager and is usually only accomplished where the forest owner or manager also garners important non-tangible benefits from those forest.

Conserving Biodiversity

Maintaining forest area is of course a good means to maintain a certain level of biodiversity. However, as can be seen from some countries, increasing forest area does not necessarily increase biodiversity nor does it necessarily mean that old forests or undisturbed forests are maintained. In many countries, reduction of net deforestation figures is at least partially due to compensatory measures, such as natural regeneration and plantations. Although depending on how these new forests are being managed, and how close they are (in time and space) to the original natural forests, these usually do not have the same species composition and biodiversity as the lost natural forests.

In terms of capacity to adapt to climate change, the change in species composition may sometimes be an advantage, if new species are better adapted to changing conditions. More problematic may be loss of diversity. Loss of diversity will make forests more vulnerable to changes, since they will not have the rich gene and species pool from which to select for the new conditions. In this respect, care should be taken of the trade-offs between mitigation and adaptation objectives; too great an emphasis on management for carbon may reduce structural and compositional diversity, thus reducing the system's inherent adaptive capacity.

A noted change in forest management in the light of climate change has therefore been an increased interest in maintaining or increasing diversity of the forests. Mixed species plantations, use of a larger number of clones and reductions in the scale of harvesting operations have been implemented as measures to maintain or increase biological diversity. These same measures are now receiving more attention because of their potential benefit in preparing forests for climate change. In addition, literature suggests that the potential yield increase from 16 appropriately selected species mixes will more than outweigh the additional costs that may be involved in mixed tree plantation establishment. The use of nitrogen fixing tree species as part of the mix, in particular in degraded lands, is beneficial for overall growth rates. Reducing the scale of harvesting operations is one way of increasing the possibility of ecological connectivity between forest patches. Plantation establishment is also an important measure that may achieve this as does planting of trees outside the forest. Although some practices are being adopted and theoretically will contribute to maintaining biodiversity, there is still a need for further research. For example, we do not yet know how much biodiversity change will cause a major and irreversible change of forest types or even ecosystems. Some authors suggest that maintaining functional diversity may be sufficient. However, it is unclear how susceptible diversity will be to climate change, if, for the different functions, the number of species that provide diversity function is strongly reduced. It is also unclear how species will react to climate change, by themselves, in combination with other species and in combination with a number of environmental and human factors. Continuous monitoring of the forests is critical to providing insight into these interactions.

Maintaining Forest Health and Vitality

The main threats to health and vitality are pests, diseases, fires and extreme weather events. In addition, diversity usually strengthens health and vitality and therefore the actions mentioned above to maintain or enhance biodiversity, also are useful for maintaining health and vitality. A number of silvicultural techniques have been developed for maintaining health and vigour of a stand. Removing old, poorly formed and damaged trees, for example, reduces the risk of spreading diseases and pests, although at the same time it may reduce diversity and thus increase the susceptibility of forests to diseases and pests.

Applying such treatments requires knowledge of the specific risks related to individual tree species and the potential benefits of maintaining poorly formed trees in the forest. Using harvesting residues on the forest floor may increase availability of nutrients for the remaining trees, thus increasing vigour, but may add to the fuel load and therefore increase fire risk. Nevertheless, timing of use (beginning of wet season) may increase the benefits and reduce the risks.

In plantation forests, a reduction of old growth and an increase in the relative presence of young stands, enhances the general health and vigour of the forest from the point of view of timber production in the medium term. Again, however, it reduces diversity, thereby reducing the adaptive capacity of the forest to externally driven changes. The decision to reduce such growth in favour of young stands needs to be taken in consideration with local conditions and management objectives.

Reducing Risk and Intensity of Damage

Reducing the risk and intensity of pests, diseases, fires and hurricane damage, along with managing the hydrological cycle, will become major concerns for many, if not all, forest managers under changing climatic conditions. Due to the complexity of measures that may have contradictory effects, there is the tendency for integrated management practices; for example, combining insect control with monitoring exercises and implementation of management practices that reduce susceptibility of the forest to insect attacks. Such practices include those treatments that help maintain the vitality of the forest, including timely thinning and species mix. For most regions however, few comprehensive management plans exist, and in most cases, plans emphasize monitoring and combating pests and diseases, rather than preventing them.

The example of pine beetle management in Central America shows how management oriented towards the prevention of fires may also reduce the risk of insect pests. Integrated fire management in forests has been promoted in many countries and by international and national agencies alike. The proceedings of a 2009 seminar in California summarize many of the experiences and achievements to date. Martell (2007) defines forest fire management as: "getting the right amount of fire to the right place at the right time at the right cost". Integrated fire management includes the following components; prediction of fire occurrence, fire prevention, fire detection, initial attack, fire management, strategic planning of resources, and fuel management. Training, knowledge sharing and planning of the different components of fire management is extremely important.

Prediction of fire occurrence requires the identification of the main causes for forest fires, followed by an analysis of factors that influence the frequency of those causes. In Mexico, this has been done through proxy indicators, such as nearness of populations and types of land use (danger assessment). This needs to be combined with characteristics of the fuel, temperatures and moisture content (risk assessment) as well as with an assessment of the value of the resource in order to be able to set priorities in fire prevention and combat activities. In many developed countries, this information is combined with daily weather forecasts to assess fire danger ratings. This has proved to be a useful early warning system.

Fire prevention is probably the most cost effective and efficient way to reduce fire damage. It requires identification of frequency, size, severity and causes of human induced fires and needs to be followed by awareness campaigns to reduce those causes or projects oriented at training and introduction of alternatives to burning forest and grazing lands. In many regions, for example Mediterranean countries and western states of the United States, watch towers are the main tool for fire detection. Patrol flights are used as well but it is difficult to plan them cost effectively. Once detected communication of location, size and burning characteristics of the fires among the fire fighters and to the public is very important, especially for the initial attack. Whereas in North America it is a problem of deciding how many air tankers to keep on standby for the initial response to fires, in many tropical countries air tankers are not available and initial attack may need to be done by local fire brigades. The quicker the response time, the smaller will be the risk of uncontrolled fires. However, such fire brigades require an extensive programme of training of the local teams, as well as sufficient equipment to be able to fight small fires. Initial attacks become more efficient if these are based on previously prepared plans that consider potential impact of fire fighting, as well as the potential risks of damage to natural, human and physical resources.

Fire management requires continuous monitoring of fuel conditions, fire behaviour and climate. With this information and short-term predictions on their changes, fire fighters can decide when and how many people to dispatch to control the fire. Fuel management is one of the more common and discussed approaches to forest fire reduction. Protection of forests has led in many cases to the accumulation of fuel, increasing the risk of forest loss, rather than reducing it. Prescribed burning is a common tool, but this needs to be supervised by experts, in order to avoid the fires to get out of control. Cutting of fire breaks is another measure often taken. These do not only form barriers to fire progress, but may also help accessibility of the forest for fire fighting crews. Technology has proved vital in monitoring and control of forest fires in regions such as North America and Europe. However, in tropical countries, where access to this technology and to the forests is more limited, fire management strategies need to include all stakeholders and all related sectors since most fires originate outside forest lands and involve non-forest stakeholders.

Risk reduction strategies for damage from hurricanes or cyclones need to consider damage to forest, agricultural crops, communities and infrastructure. Reducing risk of damage to forests can be done through silvicultural practices such as shorter rotation cycles; young trees often are more resistant to wind throw, but if thrown, will result in less biomass lost compared to larger trees. Increasing resilience, for example by maintaining good seed sources and mixed species forests, including species that readily sprout after wind throw is another means of reducing damage. In Toncontín, Honduras, the local community started to improve forest management in 1997, one year before hurricane Mitch struck. The forest is dominated by species resilient to hurricanes through different strategies including rapid and abundant seed regeneration or re-sprouting. Part of the management improvement was the retention of seed trees. Comparative studies one year after Hurricane Mitch indicated that the forest with seed tree retention was recovering its original species composition faster than forests where all commercial trees had been cut without seed tree retention.

Improving Water Regulation

Changes in precipitation patterns probably will affect forests more than changes in temperature at least over the next twenty years. The impacts will depend on the increase in atmospheric CO2 and the reaction of trees to this increase. The potential water regulation functions of forests therefore, are receiving more and more attention in the forests and climate change discussions. The role of forests in this case is vital not only as a provider of an ecosystem service (in particular in relation to the quality of water for consumption, for irrigation, for industrial use), but also for the survival of the forests themselves. Many myths regarding the function of forests in the water cycle exist. Much more research is needed and this research should include studies on the influence of individual trees and forest stands on the local water cycle and the water cycle of watersheds. As a rule of thumb, the original natural vegetation is the best way to regulate the water cycle. Planting trees in grassland swamps, such as one of the water catchment areas of Bogotá, Colombia, may result in rapid depletion of the water resources. On the other hand, leaving trees in dry areas may improve infiltration, soil structure and water holding capacity of the soils. Planting fast growing exotics in such areas may result in depletion of the soils. Dry areas, however, may also be affected by salination and in that case, care should be taken not to increase underground water levels, since this may increase the release of minerals in the upper soil layers.

Different plants have different water use efficiencies and these may change when exposed to more sunlight. Oak in Italy appears to be less sensitive to light conditions in its water use efficiency than a local Beech species. This may imply that beech stands may suffer less from reduced rainfall, since individual trees will react positively to lower stand densities while oak stands, on the other hand, may lose volume. In terms of effect on the regulation of the water cycle, oak may be easier to manage, since thinning of the oak stand will reduce water use more than in beech stands. Couralet *et al.*, (2010) found similar results for species in understory trees in Africa. Equal treatment of a forest stand in terms of modifications to the microclimate and water availability will have varying effects on the different species.

This relationship, however, needs more research. Knowledge of this correlation is useful for adaptation, species selection for local conditions and the estimation of the value of a species and its management system for mitigation. In Argentina and Chile, Nothofagus pumilio is suitable for mitigation due to its plasticity, while the regeneration capacity of N. macrocarpa was highly affected by the moisture gradient within canopy openings. Dietz *et al.*, (2006) in a study in Indonesia, found that tall trees may increase water evapotranspiration and decrease throughfall.

This suggests that smaller trees would be better conservers of water. Of course, this is dependent on how well these smaller trees allow the throughfall and stem flow water to infiltrate in the soils, and the water retention capacity of the soils. Apart from the effects on plant growth of individual species, climate change is likely to affect water availability at a watershed level. There has been an increase in the implementation of projects in upper watersheds with many positive benefits. However, the meta-analysis of Locatelli and Vignola (2009) clearly indicate that this is not always the case. Two strategies can be followed in such cases.

They demonstrated the necessity to study the effect of trees on an experimental basis, before promoting large scale regeneration or plantations. In addition, 20 monitoring of water flow at strategic points within the watershed, taking into account current land use practices will allow for changes in the planning of species composition. In most cases, at the landscape level, control of land use and forest regeneration goes beyond the control of individuals and will need some form of stakeholder collaboration to achieve the desired results.

A good case of stakeholder collaboration can be seen in the Dominican Republic, where farmers with land suitable for irrigation allowed owners of land on the ridges to occupy part of their land in return for reforestation on the ridges. It is estimated that for each hectare of irrigated land, thirty hectares are being regenerated, contributing to the availability of water for irrigation.

6

The Forest Ecological Foundation and Requirements

SCIENTIFIC DISCIPLINE OF ECOLOGY

The scientific discipline of ecology is dynamic, steeped in historical tradition but adapting to its changing environment and building its own ecological network of interactions with other disciplines. What is ecology, what have some of its high points been, and, most important, where is it going? All of us on planet earth today, not just scientists, need to know the answers. Society has become increasingly aware that we are losing crucial parts of our ecosystem, and that the activities of human beings are threatening the sustainability of the biosphere as a life-support system for humanity. As I have written elsewhere, "Ecology, the unifying science in integrating knowledge of life on our planet, has become the essential science in learning how to preserve it."

The word "ecology" did not exist when Charles Darwin published his journal, *The Voyage of the Beagle,* but one can argue that Darwin's account launched ecology as a discipline. Darwin began by reminding his readers that he had previously published volumes on coral reefs, volcanic islands, and the geology of South America. His integration of the physical and biological dimensions of the places where the *Beagle* stopped, complemented by influential contemporary work by the English naturalist Alfred Russel Wallace, defined a new and synthetic way of looking at nature—in which the patterns characteristic of particular regions found explanation in a unifying, dynamic framework. The term "oekologie" came later, the inspiration of the German scientist and physician Ernst Haeckel, but it was the theories of Darwin and Wallace that created the rich understanding that allowed the new approach to rapidly become a science.

Thus ecology sprang from a marriage between geology and natural history, rooted in observation, but answering a need for a conceptual framework. From its early roots, it became increasingly rigorous and quantitative, and a century later provided the natural discipline for responding to the environmental challenges highlighted by Rachel Carson, Paul Ehrlich, and others. Ecology and environmental studies made for a natural partnership, since the scientific study of ecology provided the tools for meeting many of the emerging environmental threats, while those threats in turn provided new stimuli for ecological investigations.

Fields like ecotoxicology, which studies the fate, transport, and effects of chemicals in the environment, and conservation biology, which seeks to preserve biodiversity, were spawned, closely linked to ecology, but much more applied in focus. The synergy among the disciplines was constructive, but it also tended to blur the distinctions in the public eye between the science of ecology and the application of ecological principles to the management of natural resources. Ecology is a scientific discipline, like physics or molecular biology, whose practitioners are driven by the search for patterns and process in nature. Their findings can inform political decisions about resource use, pollution, climate change, and other environmental issues; but advocacy regarding decisions about our environmental priorities is outside the discipline of ecology. Still, for many people, "ecologist" became a term applied to anyone who wanted to save the planet, or selected parts of it, which made no more sense than calling someone who marvels at the night sky an astronomer.

The application of ecological principles to environmental problems did not suddenly occur with the publication of *Silent Spring* in 1962; that, like ecology as a whole, had much older roots. The fishing industry provided an early case study nearly a century ago, when the distinguished Italian biologist Umberto D'Ancona puzzled over the causes of fluctuations in the fisheries of the Adriatic. Fortunately, D'Ancona was to marry the daughter of the great Italian mathematician, Vito Volterra, who had become interested in applying his skills in formal mathematics to the study of biological and social systems. Volterra is famous among mathematicians for his contributions to the theory of functionals and integral equations, but is even better and more widely known for the equations he and Alfred Lotka derived independently to describe the dynamics of interacting species. Volterra showed that the simplest of ecosystems, involving only a predator species and its prey, could go through endless cycles of changing numbers of individuals as the growth of the prey population produced a similar increase in the predator; that in turn led to a decline in the prey, followed by a decline in the predator, bringing both populations full circle.

Nevertheless, the greatest contribution Volterra made was not in his specific equations, but rather in the heterodox idea that sophisticated mathematical methods could be used to understand the dynamics of natural systems. The equations of Volterra and Lotka are taught today in virtually every first course in ecology, and the application of mathematical and computational methods has expanded into every branch of ecology, giving us, for example, integrated models that deal with the interaction between a changing climate and the growth of forests and other vegetation. The influence of mathematics and computation in biology as a whole has also expanded in the last decade, and fields like systems biology (the study of the interplay among the parts of biological systems, like molecular or metabolic systems) and computational biology (with its use of mathematics and computation) have become essential parts of any modern biology department. It should not be lost, however, that through the con- tributions of Volterra and those who followed in his footsteps, ecology was the first subdiscipline of biology to become deeply quantitative.

That is important, because fisheries and forests are perhaps the most important cases of natural populations exploited by human beings for food, fiber, and fuel; they are examples of what today are termed "ecosystem goods and services." Managing fisheries raises unique challenges as compared with agriculture, or even to forests, because so much of the ocean's resources are beyond the control of the manager. The economic importance of fisheries, and the importance of managing them to optimize the benefits human beings derive from them, has led to a rich theoretical and applied literature, with some successes and many failures. Marine fisheries have always provided crucial sustenance to human beings, but many of the most favored stocks have collapsed because of overfishing. That has led to recognition that we need to pay more attention to the insights of D'Ancona and Volterra, and to manage fisheries as components of complex ecosystems rather than as independent entities. Even today, most management efforts are focused on particular species, like salmon, for many of the same reasons that the Endangered Species Act refers to individual species and populations; such specificity may reflect the constituencies that salmon or trout, or pandas or tigers, enjoy—and more generally simplifies management. There is increasing recognition, however, that any species is part of a larger network of species, competing for common resources, and that such interconnections cannot be ignored any more than economic policy makers in the United States can ignore the problems of Iceland or Greece.

Fisheries provided only one early and important example of the application of mathematical and ecological perspectives to human welfare. Two other examples come from epidemiology and demography. Infectious diseases involve a fundamental ecological interaction, that between a parasite and its host. Parasite-host systems are similar in many ways to predator-prey systems, and hence exhibit the same potential for oscillation that Volterra demonstrated. Such oscillatory dynamics are only too familiar to us, for example, in the seasonal fluctuations of influenza cases. There is a difference, however, in that parasites live within hosts, and so may die when their hosts die unless they can be transmitted to other hosts. From an evolutionary perspective, that suggests that parasite-host systems will evolve to favor less virulent parasites, because the most virulent strains kill their hosts before they can spread to other hosts;

there is considerable evidence to support that concept. One remarkable example involves documentation by the virologist Frank J. Fenner, beginning in the 1940s and 50s, of the evolution of reduced virulence in the myxoma virus in Australia when it was introduced to control the European rabbit; again an ecological perspective proved crucial to successful management of the plague of rabbits.

Demography provided perhaps the earliest example of the application of both ecological and mathematical thinking to problems of natural populations. Demographers like John Graunt, as early as the 17th century, used sophisticated mathematical methods to project human population growth, building on methods that were known for centuries before, tracing back to the mathematician known as Fibonacci. Graunt also recognized that the observed growth rates for the population of London were not sustainable, and could not have been in place historically, or else the population of London would have been much larger than it was. In the early 18th century, Thomas Malthus, in *An Essay on the Principle of Population,* emphasized that essentially ecological factors, like famine and disease, ultimately restrain the growth of population; that insight was incorporated into a number of mathematical modifications of the basic equations of exponential growth. The science of demography has evolved an elegant mathematical underpinning since then, and its evolutionary dimension—life-history theory, which seeks to elucidate the impact of natural selection on the life cycle, is a cornerstone of ecological theory.

Of course, because of the implications for economic growth, these issues will always remain controversial. In 1968, Ehrlich's *The Population Bomb* warned about the devastating possibilities, like mass starvation, of unlimited population growth. That has led to spirited and often rancorous debate about the inevitability of population limitation and the potential for technology to fill the needs of a world population growing without limits. To an ecologist, it seems evident that earth's resources are ultimately finite, and that populations cannot grow without bound. Where those limits are, and how large a population the world can sustain, is the essence of the debate.

And where are we now? The writings of Darwin and Wallace clearly influenced the development of ecology; but perhaps even more fundamentally, they also launched the field of evolution. Natural selection is just one of many factors influencing evolutionary change, but it is the most important in providing a link to the ecological context within which organisms exist and interact. When Darwin developed his theories, our understanding of the genetic mechanisms underlying evolution was in its infancy. In the early parts of the 20th century, the theory of population genetics developed rapidly, in close partnership with animal and plant breeding, and with a strong mathematical foundation. Then in the middle of the 20th century, unification took place between population genetics and evolutionary theory. The Modern Synthesis provided a framework for a similar unification with ecological studies, but that step had to wait as genetics turned to developing a deeper understanding of the molecular details of genetic control of basic aspects such as development and physiology. As that foundation has become solidified, increased attention has become directed to systems biology—how the complex assemblage of genetic circuits cooperate to produce and run the organism, how the organism interacts with other organisms and its environment, and how whole communities and ecosystems emerge from evolutionary processes at the microscopic level. The collection of vast amounts of "metagenomic" information, involving genetic material of many species in the same environments, holds great potential for allowing us to extend evolutionary studies beyond individual species, to the complex ecosystems in which fish, pandas, and tigers live.

Perhaps the scientist who most influenced the development of ecology in the last century was the limnologist G. Evelyn Hutchinson, who studied inland waters but wrote trenchantly on almost every aspect of ecology, providing an evolutionary context for understanding ecological characteristics and ecosystem patterns. However, although the unification of the multiple levels of organization may have been clear in Hutchinson's mind, three or four decades ago the subjects of population biology and ecosystems science were worlds apart intellectually, and leaders of both struggled to find ways to communicate with each other. That chasm has narrowed dramatically, as population biologists have begun to emphasize the importance of coevolution—how evolving populations influence one another—and niche construction—how organisms shape their own ecological roles and those of other organisms. Ecosystem scientists now strive to understand patterns of nutrient use and nutrient cycling within an evolutionary framework, epidemiologists focus on the coevolution of hosts and parasites and seek understanding of the evolutionary history of viruses and bacteria, and rapid advances in genomics and metagenomics raise challenges for understanding their relevance for the distribution and abundance of organisms in their natural environments.

The scientific discipline of ecology has thus been, in some sense, a key node in an ecology of scientific disciplines. From its roots in natural history, it has built partnerships with botany and zoology; with geology and paleontology; with mathematics; and with evolutionary studies, from molecular biology to population genetics and development. It has not ignored physics and engineering along the way, in its need to understand the mechanics of how animals and plants grow and move, or how they capture energy and resist stress—why, for example, trees are shaped the way they are. Ecologists make their living by recognizing the interconnectedness of different parts, and different disciplines.

But what is missing in the story to this point are strong links to the social and policy sciences, and to the humanities. Those dimensions have not been entirely missing in the past. Some conservation ecologists have recognized the importance of making clear that biological diversity must be preserved not just for its own sake, but because it underlies a wide range of goods and services that are essential to all life on our planet, including food, fiber, fuel, and pharmaceuticals. We also receive indirect benefits from pollination, from plants that moderate climate change by taking up carbon dioxide, and from habitats that sequester nutrients and toxins; all of those are active topics of ecological investigation today.

What we must do now, however, is to unify science and the social sciences and humanities further in the service of preserving the earth. That challenge is stretching ecology in new directions. To sustain the planet, we must be able to protect the goods and services ecosystems provide, and that requires knowing what they mean to us. In large part, that is the domain of economics. Bringing ecology and economics together is not new, and most land-grant universities have strong departments of agricultural and resource economics that deal with the economics of agriculture, fisheries, and forests. But new dimensions in environmental and ecological economics have been developing rapidly, focusing on questions about the value of whole ecosystems, about the optimal design of nature reserves, and about how to manage public goods. Broadly understood, environmental public goods include the air we breathe and the water we drink, the fish we catch and the lakes and oceans that sustain them, as well as the antibiotics that have contributed so much to the advancement of medical care.

Still, cooperation for the common good is vulnerable to cheaters who can derive benefits without paying their fair share, and it cannot be sustained without new mechanisms and institutions. Our political leaders have already begun to recognize that in dealing with the world's interconnected financial systems; it applies with at least equal force to our environment. Solutions to the great environmental challenges we face will require deep partnerships between ecologists, economists, and other social scientists.

Scientists will also need to work with humanists and ethicists, as we deal with the core problems of intergenerational and intragenerational equity, and the powerful role of social norms in shaping individual behaviors. How much of our resources is it equitable for us to consume today, and how much should we leave for future generations? How do we deal with the growing inequity in the distribution of wealth within our populations today, and the implications for access to natural resources?

As with the partnerships ecology has forged in the past, the benefits are likely to be mutual; indeed, months before the current financial crisis, my co-authors and I suggested in an article in*Nature* that an ecological perspective would be helpful in dealing with issues of systemic risk in the banking system because, we argued, the system was too interconnected and vulnerable to contagious spread of disturbances due, for example, to the subprime-mortgage mess. The banking system, just like the economy, is a kind of ecological system, and it should be managed as one.

Ecology views biological systems as wholes, not as independent parts, while seeking to elucidate how the wholes emerge from and affect the parts. Increasingly, such a holistic perspective, rechristened at places like the Santa Fe Institute as "the theory of complex adaptive systems," has informed understanding and improved management of economic and financial systems, social systems, complex materials, and even physiology and medicine. Essentially, that means little more than taking an ecological approach to such systems.

RISE OF ECOLOGY AS SCIENCE

The first author of the term "ecology" is uncertain, though several writers used it in the mid-nineteenth century. Many biologists grant credit to the Germany biologist Ernst Haeckel, who used the term in 1866 to refer to the interrelationships of living organisms and their environment. A few years earlier, in 1857, Henry David Thoreau also spoke of ecology among other fields of biology and natural history, but he apparently did not provide a definition of the term. In any case, the origins of ecologic knowledge are much older than either Haeckel or Thoreau, and date to the origins of biology itself even though the word was not used until the nineteenth century.

THE ORIGINS OF BIOLOGY

The beginnings of ecologic knowledge for modern man lay in the development of biology and medicine as scientific disciplines. The fourth century Greek, Hippocrates, often called the "Father of Medicine," emphasized certain ecologic factors in the genesis of health and disease.

Aristotle, sometimes referred to as the "Father of Biology," classified animals according to their habits and habitats; *e.g.*, gregarious vs. solitary, carnivorous vs. herbivorous, resident vs. migratory, etc. Although he was a good naturalist he could not yet be considered an ecologist for he did not really consider interrelations of plants, animals and their environments other than in the most superficial terms. Theophrastus, a student of Aristotle, was regarded by some scholars as the first ecologist, though it should be emphasized that the word ecology was not used until much later. Nonetheless, Theophrastus discussed plants and their environments in a moderately systematic way. He studied plant types and form in relation to altitude, moisture and light exposure.

About the time of Christ, the Hebrews had some ecological notions, but of a very general and rather obvious sort.

In Rome, Pliny the Elder was a naturalist with a system of classification somewhat like that of Aristotle; that is; he classified plants and animals by their habits and habitats. He was a great compiler and cataloger of facts, but most of this was basic natural history and could not be considered truly ecological in orientation.

Beginning slowly in the twelfth century and accelerating rapidly in the sixteenth century, natural history flourished as a great scholarly endeavor. The description and cataloging of animals became a favorite scientific activity. Albertus Magnus was perhaps the leading naturalist of his time who wrote on plants and their environments. In the sixteenth century, Konrad Gesner, Aldrovandi, and Cordus all began to relate biology and natural history to geography and chemistry even though the latter had barely begun as a science.

It remained for Robert Boyle, sometimes called the first modern chemist, to relate the origins of chemistry to vital processes in plants and animals. He did experiments on the effects of low air pressure on a variety of animals from frogs to chickens and mice. In one series of experiments he showed the ability of ducks to withstand a shortage of air longer than a chicken.

From the late seventeenth century through the nineteenth century, five major fields of scholarship laid the foundation for the development of ecology as a self-conscious scientific discipline in the twentieth century. These five fields were: (1) natural history and faunal exploration, (2) environmental physiology and responses, (3) evolution and natural selection, (4) population studies and demography, and (5) ecological geography and conservation.

These did not develop as independent subjects, of course. Several men played key roles in the origins of two or more areas. For example, Charles Darwin was a naturalist of the mid-nineteenth century who made outstanding contributions to natural history, faunal exploration, and evolution. Reaumur and Buffon in the early and mid-eighteenth century both made contributions in natural history and environmental physiology. Thomas Malthus provided the stimulus for population studies, and he also laid the foundation for Darwin's work in natural selection. Although these and many other interlocking forces were active in the development of ecology, it is helpful to consider them sepaecology. More complete information on any of these trends can be found in books by Allee, Singer and Nordenskiold.

EXPLORATION OF NATURAL HISTORY

Man's interest in natural history has been a primary motivation and developmental source of ecology since the time of Aristotle. Some of the major naturalists of the Renaissance who contributed to this development are noted below.

Rene Reaumur was a French naturalist and physicist who wrote a six volume work entitled, *Memoires pour Servir a l'Histoire des Insectes*. This work discussed the life habits and community patterns of social insects. Reaumur also investigated the natural history of crustaceans and molluscs.

Carolus Linnaeus or Karl Yon Linne was the first great modern systematist. His classic work entitled *Systemmae Naturae* established the binomial system of nomenclature which is the standard scientific method for naming organisms. All modern scientific names followed by a capital L were first applied by Linnaeus.

He also furthered the sciences of plant and animal geography and studied plant associations in different regions. George Louis Leclerc Buffon, another eminent French naturalist, wrote the forty-four volume *Histoire Naturelle, Generale et Particuliere*, a vast compendium of most of the natural history knowledge of the day. He was interested in the adaptations of plants and animals to their environments. He brought together a great deal of information about climate as a variable, compelling modification of plants and animals. In some ways his writings anticipated the evolutionary ideas of Darwin, but they were not fully expressed or developed. He definitely opposed Linnaeus, ideas of the fixity cf species, and he massed considerable evidence demonstrating variation and mutability in living organisms.

In England, an outstanding naturalist of this time was Gilbert White. His book, *The Natural History of Selbourne*, still provides thoughtful and enjoyable reading about the natural history of southern England.

Nineteenth century England produced a golden era for natural history. One of the leading British naturalists of this period was Alfred Russell Wallace who wrote *Island Life*, and several other leading books about the distribution of plants and animals of the Malay archipelago. Wallace also achieved a theory of evolution at the same time as Darwin, and this independent development of a great biological principle represents one of the fascinating chapters in the history of biology. More will be said of this later. Charles Darwin was an amazingly productive scholar who wrote more than a dozen outstanding books and monographs on a variety of biological topics from 1842, when he published *Coral Reefs* to 1881 with the publication works, of course, were the *Origin of Species*, and the *Descent of Man*. With these, especially the former, he presented his great interpretive theory of natural selection and evolution.

Other nineteenth century English naturalists of considerable influence were H. W. Bates, who wrote *Naturalist* on the Amazon in 1863; Thomas Henry Huxley, the great physician, anatomist and zoologist, who became one of Darwin's staunchest supporters; and Edward Forbes, a marine biologist who was especially interested in the dynamic relations between marine organisms and their environment.

Before the time of Darwin and Huxley, James J. Audubon in America was compiling his monumental *Birds of America*, and shortly thereafter Louis Agassiz, the Swiss-American naturalist, compiled four volumes of *Bibliographia Zoologica*, along with other books on fossil fishes, glacial activity and naturalistic expeditions to Brazil.

Toward the end of the nineteenth century, Henri Fabre in France, published ten volumes on the *Life of Insects*, and A. E. Brehm of Germany was compiling material for the subsequent publication of the thirteen volume, *Tierleben*. Other influential scholars of the nineteenth century were C. Hart Merriam, who devised a classification of major life zones—a system of habitat classification still in wide use, and S. A. Forbes, who wrote a series of papers on the concept of ecological communities. This began the period of the self-conscious rise of ecology as a scientific discipline, and we shall return to this topic after brief consideration of other trends in the eighteenth and nineteenth centuries.

ENVIRONMENTAL PHYSIOLOGY

Rene Reaumur and George Buffon were primarily naturalists of the eighteenth century, but they were also interested in environmental responses to temperature, light and altitude, and they may be considered early students of environmental physiology. This field did not really gain momentum until the nineteenth century, however, when Justus Liebig developed his analysis of limiting factors. He was interested primarily in physical factors which limit the distribution and abundance of plant and animal life. We recognize his basic contribution today in Liebig's Law of the Minimum, which states that an organism will be limited by that factor most closely approaching its minimum life support requirements. Also at this time, Jacques Quetelet studied dormancy and hibernation in plants and animals in relation to temperature and shortly thereafter, Alphonse de Candolle studied plant germination in relation to temperature and moisture.

Semper wrote an important ecological treatise in 1881 entitled, *Animal Life as Affected by the Natural Conditions of Existence*, in which he established the principle of food chains and what we now call a "pyramid of numbers".1I He was interested in predator-prey interactions, protective coloration, and the spatial relations of animals. He showed that pond snails and fresh water isopods would be stunted if crowded in a small volume of water, and he postulated the existence of chemical factors passing between animals in the environ-ment and influencing growth and development. We now recognize these as "ectocrines" or environmental hormones, and they represent major area of current research. This was a fairly sophisticated concept in the 1800's.

Davenport continued this line of investigation and wrote two volumes in 1897 and 1899 entitled, *Experimental Morphology*. He studied animal behavior and development in a variety of artificial environments and media, and he analyzed a wide range of environmental responses: chemotaxes, hydrotaxes, geotaxes, phototaxes, thermotaxes, electrotaxes, and so forth. He also demonstrated the viability of frozen cells, and thereby opened up several lines of study and practical utilization. Semper and Davenport gave strong impetus to a branch of ecology which is still at the forefront of modern ecology, the interrelations of physiology and ecology.

EVOLUTION

The history of evolutionary thought is closely interwoven with that of natural history, and most of the leading figures in developing theories of evolution were the great naturalists of the eighteenth and nineteenth centuries. George Buffon and Jean Baptiste de Lamarck were both environmental evolutionists of the eighteenth century and early nineteenth century respectively. That is, they believed that plants and animals were changed by environmental influence. Lamarck is most famous, of course, for his *Theory of Acquired Characteristics*, which we now know contains some major misconceptions about the mechanism of evolution. Some 60 years before Lamarck, Buffon was developing somewhat similar ideas of evolutionary change in his *Histoire Naturelle* of 1749:

If we again consider each species in different climates we shall find obvious varieties both as regards size and form; all are influenced more or less strongly by the climate. These changes only take place slowly and imperceptibly; the great workman of Nature is Time: he walks always with even strides, uniform and regular, he does nothing by leaps; but by degrees, by gradations, by succession, he does everything; and these changes, at first imperceptible, little by little become evident and express.

These writings of Buffon strongly influenced Lamarck in two major books, *Philosophie Zoologique*, in 1809 and *Histoire Naturelle des Animaux sans Vertebres*, in 1815. In these works he presented his theory of evolution, based on the inheritance of acquired characteristics, sometimes referred to as the "use and disuse theory." He believed, for example, that the giraffe's neck had become long because he continually stretched higher for leaves to feed on, and thus each generation was born with a slightly longer neck as a result of stretching. Similarly, he believed that blind cave fish lost their eyes because they failed to use them. Although we can recognize elements of truth in these theories, we now know that Lamarck's understanding of the mechanism was incorrect. We know that environmental changes affecting only the somatic cells of the body (in contrast to the germinal or reproductive cells) are not transmitted to subsequent generations. The evolutionary theories of Darwin and Wallace emerging about 40 years later were similarly ecological in emphasis, but radically different in how evolution occurred. They developed their views independently and almost concurrently, placing emphasis on population pressure and struggle for existence as the driving forces of evolutionary change. They envisaged survival of the fittest and natural selection as the mechanism of evolution, working on the foundation of inherited variations in plants and animals. They knew nothing of genetic mutation, but with this knowledge added some years later, their central ideals of natural selection remain basically intact at the present time.

Darwin's and Wallace's initial papers were published simultaneously in 1858, and this provided an unusually interesting footnote for the history of biology.

POPULATION STUDIES

From the sixteenth century onward, some scholars have worried about population growth and pressures. Machiavelli, Botero, Buffon, and Franklin were all scientists who partially anticipated Malthus by expressing concern about population growth, starvation, disease and war. Yet Malthus, of course, was primarily responsible for focusing controversial attention on these topics.

Although we often remember him as the Reverend Thomas Robert Malthus, he was, for most of his life, a Professor of History and Political Economy at Haileyburg College. While still a curate in Surrey, however, he published his controversial book, *An Essay on Population*. In this expansive volume he observed that human populations tend to increase geometrically and thus exceed the means of subsistence which increase arithmetically. He felt that unless populations were controlled by voluntary constraint, they surpassed their level of support and were then decimated by starvation, disease, war and poverty.

This theory produced a storm of protest, and it has remained controversial to modern times. Generally, it fell into some scientific disrepute, for many scholars believed that Malthus failed to recognise the exact potential of modern science in overcoming famine and disease. Many felt it painted too dismal a view of human potential. Thus, 30 years ago there were few scholars who actively endorsed the views of Malthus, and many books were written to refute them.

More recently Malthusian doctrine has come back into favor, and it is being taken more seriously now than in the 1930's. Many writers who grant the potential of science and technology to solve the problems of population pressure, admit that the Malthusian fears of starvation, war and disease still haunt the earth. In future sections, more will be said of this, and we will spend considerably more time on the history of population study and demography.

Incidentally, it has been an odd quirk of scientific history that Darwin was favorably influenced by Malthus, and relied heavily on his doctrine to establish his theories of natural selection. Subsequently, Darwin's theories were accepted in scientific circles long before those of Malthus.

Population studies resumed great importance in the twentieth century with the work of Raymond Pearl, Thomas Park, Charles Elton and many others. These studies all form part of the modern and self-conscious development of ecology.

NUTRITIONAL AND ECOLOGICAL REQUIREMENTS

For its growth and for other life processes, the tree requires food and like other green plants it builds up its food materials from simple substances. The chemical constituents of a tree are chiefly carbon, hydrogen and oxygen, together with some nitrogen and certain minerals such as potash, lime, iron and phosphorus. The carbon is obtained from the small amount of carbon dioxide gas that is always present in the atmosphere and enters the plant through small openings in the leaf surface known as "stomata." Oxygen and hydrogen are obtained from the soil through the root system in the form of water carrying nitrogen (nitrates) and mineral nutrients in solution. The synthesizing of these elements to form substances that the tree requires -in particular the combination of carbon, hydrogen and oxygen to form carbohydrates — takes place mainly in the leaves in the presence of the characteristically green matter (chlorophyll) and of sunlight, which supplies the necessary energy. Air, water, chlorophyll and light are therefore essential to the production of the carbohydrates-in particular cellulose and lignin -which form by far the greater portion of the tree. Soil moisture is especially important, as a deficiency of water inhibits photosynthesis. Tree species differ markedly in their optimum water requirements and also in their sensitivity to departures from the optimum.

Tree species also differ greatly in the amount of light they require, or, alternatively, in the degree of shade they will tolerate. It is customary to classify species as shade-enduring or tolerant (of shade) and light-demanding or intolerant (of shade), but for most species the younger the tree and the better the soil quality on which it is growing, the greater its tolerance of shade.

The environmental factors important in tree growth — solar radiation, air temperature, humidity, precipitation and the physical and chemical nature of the soil — are usually spoken of as the factors of site or locality and their total effect with reference to a species as the site quality. Site quality is evaluated either directly by examination and classification of soil and other factors, or indirectly, as for example, by using ground vegetation as an indicator of soil conditions, or by relating the height of dominant trees to the age of the stand.

Since the bulk of the tree is derived from carbon dioxide and water, a relatively small demand is made on the soil in growing a crop of timber; thus land too poor for profitable cultivation may be capable of growing a crop of timber satisfactorily. Deeply penetrating roots bring up from the deeper soil strata mineral nutrients that would not otherwise be available; a relatively high proportion of these minerals may be deposited annually on the surface soil through the falling of leaves, twigs and other debris.

The leaves, which play such an important part in the nutrition of the tree, function for only a limited time before dying. Deciduous trees lose all their leaves at the end of the growing season, whereas evergreens lose only a portion of the older leaves each year.

The growth and general well-being of a tree depend primarily on soil and weather conditions. Also of vital importance to the future of each member of the forest community is the intense competition that develops for space and food requirements-a struggle that inevitably kills many of the competitors and brings about significant changes in the environment of the survivors.

LONGEVITY

Under favorable conditions a tree usually grows vigorously until it reaches maturity, after which the growth rate gradually slows down. Overmature trees grow little and show signs of decadence. The normal life expectancy of trees varies widely with the species: some temperate and tropical hardwoods reach maturity in less than 100 years and some softwoods in 100 to 200 years, but many species of both softwoods and hardwoods require 200 to 500 years to mature. On the Pacific coast of North America, ages of 500 to 700 years are common; the giant sequoia in California has been known to live for over 3,000 years, an example of extreme longevity.

Resistance to Damage

Trees, like other organisms, are damaged and destroyed by various agencies, including fire, insects, fungi, grazing animals, industrial fumes and such climatic factors as wind, frost and snow. The capacity to resist such damage depends upon the vitality and the protective devices of the species. A knowledge of the relative resistance of different tree species is of importance in forest management.

Distribution

Apart from the chance results of geological history, the continental distribution of tree species and the limits of tree growth, both in altitude and latitude, depend primarily on heat; the east-west distribution, on the other hand, is more often determined by moisture conditions—the rainfall in the interior of a continent may be too scanty to support tree growth. Thus each species of tree has a range of natural distribution determined primarily by climate. Obviously there is a great deal of overlapping and many species grow under similar conditions — in fact it is common to have a number of species sharing the same area in a mixed stand.

Pure stands of a single species do not commonly occur in the natural forest, though stands approaching this condition may be found where a species is well adapted to extreme conditions of climate or soil moisture, or where fire has brought about a situation particularly favorable to one species. The most heterogeneous and complex forest communities are found in tropical regions of high rainfall.

The distribution of a species within its natural range is determined mainly by soil, topography and competing vegetation, as well as by local variations in climate, which also affect the long-term development of many species: climatic races or strains develop, more or less adapted to local variations of climate within the range as a whole. This fact accounts, in part at least, for the wide variation in results obtained by using seed of other than local origin in establishing forests.

Scots pine, for example, has developed a number of climatic races and experience in both Europe and North America shows that efforts to introduce a "foreign" race of this species without a proper correlation of climatic conditions at seed source and planting site will almost certainly prove unsatisfactory, particularly in the susceptibility of the trees to damage by snow and sleet and the incidence of disease. This, of course, also holds true for attempts to establish any tree as an exotic outside the boundaries of its natural range. However, where the climatic conditions of two regions are similar, the introduction of tree species from the one to the other may be undertaken successfully, as, for example, in the introduction of conifers from the Pacific coast of North America into Great Britain.

An understanding of the various biological factors discussed above is of fundamental importance in the development of forestry practice, particularly in the fields of silviculture and protection.

It must be emphasized, however, that practical forest management must depend also on the knowledge and application of sound business principles and on the development of suitable methods and techniques to be used in collecting, analyzing and organizing the basic data to be obtained from the forest.

Branches of Forestry

As is implied by what has been said above, forestry has a number of subdivisions, each of which is concerned with some particular aspect or phase of the subject. Hence, an understanding of forestry itself is dependent on knowledge of its parts, the extent to which they are interdependent and how best they may be co-ordinated.

VERTEBRATE-MEDIATED ECOLOGICAL PROCESSES

Ecological interactions have critical importance in the maintenance of biodiversity and ecological functions such as productivity and turnover. Mutualistic interactions are extremely common and include interactions between all major organismal groups, they have been largely neglected in the ecological literature, with greater emphasis placed on competition and predation. Many interactions may be quite unspecific with a great degree of ecological redundancy, such that the loss of a single species member of a mutualism results in little change in the overall outcome of the interaction. The extent of redundancy and the resistance of ecological interactions to anthropogenic disturbances are poorly known, yet of critical importance to forest managers in determining extractive mechanisms and limits. Vertebrate-plant interactions should be of interest to both forest managers and conservation biologists owing to the roles of vertebrates in plant recruitment and in regulating forest structure and composition through pollination, seed dispersal, and seed predation and herbivory.

POLLINATION

Pollination by vertebrates is almost non-existent in northern temperate forests where most species are wind-pollinated. Indeed there is a trend of increasing wind pollination with increasing latitude. In tropical forests, high tree diversity and wide spacing between conspecifics limit the efficiency of wind pollination and tree species in lowland tropical forests are almost all pollinated by animals (including insects). To attract vertebrate pollinators plants need to produce nectar in quantities much greater than that required for insect pollination. The energy costs of nectar production to plants can therefore be considerable. Thus nectar produced by vertebrate-pollinated plants tends to be relatively dilute. Vertebrate-pollinated plants are therefore most abundant in wet tropical climates where water does not limit nectar production, and are rarely found in dry forest or savannah habitats. Among vertebrates, bats, birds and a few non-flying mammals are the only known pollinators; although they pollinate only a minority of plants, they remain crucial for the successful reproduction of many large forest trees and a much greater variety of forest understorey plants.

Well-documented cases of non-flying mammal pollination involve 59 mammal species distributed among 19 families and six orders. The marsupials of Australia (12 species) and primates of Madagascar (9 species) and South America (10 species) are the best-known examples. However, the importance to plants of non-flying mammals as pollinators is debatable, as many of these plants are often served by other vertebrate and insect pollinators. Only in Madagascar do non-flying mammals (lemurs) act as sole pollinators of the plants they visit, and the precipitous decline of some lemur species may cause an associated decline of the plants they pollinate (*e.g.* the traveller's tree *Ravenala madagascarien-sis* pollinated by the endangered black-and-white ruffed lemur). Selective exclusion experiments have demonstrated that marsupials are at least as effective as birds in pollinating *Banksia* in Australia and *Protea* spp. in South Africa; indeed, declining small-mammal activity at *Banksia* inflorescences due to habitat fragmentation resulted in a sharp decrease in seed set of *Banksia goodii.* However, documented cases of reduced seed set due to declines in mammal pollinators are rare. In most cases, evidence for effective pollination by non-flying mammals is largely circumstantial and, while non-flying mammals may feed at flowers and occasionally pollinate them, this service must be balanced with the damage they cause.

Bats pollinate species belonging to several tropical plant families including Bombacaceae *(Pseudobombax),* Mimosaceae *(Parkia),* Passi-floraceae *(Passiflora)* and Caesalpiniaceae *(Bauhinia);* on oceanic islands, such as Samoa, the majority of the dominant canopy tree species may be pollinated by bats. A number of economically important tropical trees rely on bats for pollination, including durian, banana, neem and timber species of *Eucalyptus* and several species of palm. Although it is energetically expensive to produce bat-attracting flowers rich in nectar, plants benefit from extensive pollen transfer as bats forage over large areas. Habitat change, hunting and the introduction of exotic species is causing the decline of bats across the tropics and several species are threatened with extinction. This is of particular concern on oceanic islands, where alternative pollinators for endemic tree species are rare or absent.

In the Neo-tropics hummingbird pollination is widespread among several plant families, while old-world pollinators include a much greater variety of birds, primarily sunbirds but also leafbirds, lorikeets, white-eyes and honey-eaters. However, birds are rarely the exclusive pollinators of plants and their importance relative to insect pollinators is not well resolved. Declines in fruit set due to temporary absence of hummingbirds have been noted, even where other invertebrate pollinators were present, but obligate pollinator dependency by plants on hummingbirds or sun-birds has not been documented. Rather, these birds may be important as biological 'insurance' where temporal fluctuation of invertebrate pollinator populations may be severe.

Loss of vertebrate pollinators resulting from forest conversion or fragmentation is only likely to affect plant seed production if the plant-pollinator interaction is highly specialized. Most plants are pollinated by a suite of mainly general-ist pollinators and are unlikely to suffer a large decline with the loss of one or a few closely related pollinator species. Nevertheless, in tropical floras specialist interactions among animals and plants do exist. On the small isolated Caribbean island of Tobago the absence of hermit hummingbirds resulted in reduced seed production of *Mandevilla hirsuta* (Apocynaceae) compared with the neighbouring island of Trinidad where the birds were present. In contrast, there was no difference across islands in seed production of *Justica secunda* (Acanthaceae) that is pollinated by a more general array of hummingbirds. In fact most hummingbirds appear to be quite plastic in their habitat requirements and regularly use edges and move between primary and secondary habitats to take advantage of a wide variety of floral resources; there is also evidence that sunbirds, which in the Palaeotropics fill a similar ecological niche to hummingbirds, are more abundant at flowering trees in disturbed secondary forests. However, predicting the impacts of habitat and landscape modification on plant-pollinator interactions is difficult as both the behaviour and abundance of pollinators may change as a result of changing resource distribution. Pollinators may also depend on resources that are far from the plants they pollinate. In peninsular Malaysia, for example, the fruit bat *Eonycteris spelea* is the exclusive pollinator of durian trees but feeds primarily in coastal mangroves. The bats visit, and pollinate, durian *en route* to the mangroves from their limestone-cave roosts. Destruction of either the caves or the mangroves could affect the bat population and hence the pollination of durian.

Seed Predation

Predation of seeds by animals has been proposed as a mechanism for the maintenance of the high tree species diversity in tropical forests. This theory suggests that density-dependent consumption of seeds by mammalian seed predators prevents domination of the vegetation by species that have competitively superior seeds or greatest seed production. Terborgh (1988) suggested that vertebrate seed predators selectively prey upon the largest tree seeds, and where carnivores are eliminated following perturbations forest composition might shift to smaller-seeded species. There is some evidence to support this prediction from forests in Panama and Mexico. In Panama, high densities of small mammals on Barro Colorado Island due to an absence of large predators is thought to be the cause of recruitment failure among the large canopy tree *Dipteryx panamensis* due to the seed predation pressure exerted by the small mammals.

Some studies show seed predation by vertebrates (most often rodents) to be higher in mature forest compared with early successional forest, while other studies found the opposite. Rodents tend to avoid large areas of open forest or pasture probably due to the lack of protective cover and the scarcity of food resources. However, seed predation intensity under different disturbance regimes varies greatly with the species of seed plant making it difficult to draw generalizations to guide forest restoration programmes.

However, where seed predator densities are high, commercially available low-value grains might be used to satiate the seed predators in order to minimize mortality of the desired species. Use of these methods should be based on a sound knowledge of the factors that limit the population dynamics of the predator species, otherwise the availability of superabundant food resources leads to rapid population growth and subsequent heavy seed mortality in later years.

Seed Dispersal

Seed dispersal has been favoured by evolution because it decreases the probability of seed predation, which is often disproportionately high close to parent trees, and increases the probability that seeds find a suitable establishment site.

Thus seeds have become adapted for dispersal by a variety of vectors, the most common of which are wind and animals, and the role of animals in the dispersal of seed and fruit has received much attention in the biological literature. Seeds carried by animals often have edible appendages (*e.g. Acacia* spp., *Anacardium* spp.) or are enclosed within fleshy fruits that are consumed and the seeds ejected later. In some cases the seeds themselves are harvested but dropped, or stored and later forgotten or abandoned.

In Neo-tropical forests, 50-90 per cent of the canopy trees and nearly all shrubs and subcanopy trees bear fruit adapted for animal dispersal, while a smaller but still substantial proportion of species in Palaeotropical forests are also dispersed by animals (35-48 per cent and 70-80 per cent, respectively). Birds, bats and scatter-hoarding rodents are the major dispersers of tropical tree seeds, while other seed dispersers include primates, fish in Amazonian forest, tortoises, some herbivorous lizards and even a frog.

In temperate coniferous and deciduous forests, seeds may be more often dispersed by wind or by cache-hoarding mammals and birds. However, this dichotomy between temperate and tropical forest dispersal mechanisms has been challenged by Wheelwright (1988), who asserts that interactions between plants and their fruit-eating bird dispersers are as complex and ecologically important in temperate regions as they are in the tropics. Howe and Smallwood (1982) state that more than 60 per cent of temperate trees have animal-dispersed seeds and record a significant negative correlation between annual precipitation and the percentage of wind dispersal. Nevertheless, there are more species of fruit-eating birds in the tropics and they comprise a larger fraction of the avian biomass. Only in the tropics are there bird species that rely exclusively on a diet of fruit (*e.g.* fruit pigeons *Ptilinopus* spp. in Australia and *Treron* spp. In South-East Asia), and most birds with a predominantly fruit diet are tropical species.

Seed dispersal syndromes are thought to reflect coevolutionary trends between plants and vertebrate (bird and mammal) seed dispersers. Bird-dispersed fruit are small and red, black, blue or purple, are conspicuous by sight rather than smell, and persist on the tree until removed by a disperser. Mammal-dispersed fleshy fruits are large, green or brown, emit odours, and frequently abscise shortly after ripening. However, the existence of clearly defined seed dispersal syndromes has recently been questioned and consumption of a particular type of fruit by an animal does not imply that the fruit is adapted for dispersal by that animal. The emerging view is that most interactions between plants and avian seed dispersers, in both tropical and temperate regions, are generalist, opportunistic and largely inefficient (from the plant's perspective), and there are few examples of tightly coupled coevolved mutu-alistic interactions. One well-documented exception is the commercially important brazilnut *Bertholletia excelsa,* which is entirely dependent on agoutis to gnaw open the seed capsules and release the seeds.

Animal behaviour can be crucial to the success of plant dispersal and recruitment. Animal seed dispersers often place seeds in sites suitable for germination, and seeds of some species (*e.g.* oak) will not germinate unless they are buried. Burial further reduces the probability that the seeds will be found by seed predators. Bats may reduce the suitability of fruit to insect seed predator attack, leading to increased survival of seeds dropped at a bat roost.

Animal digging, burrowing and tunnelling (by armadillos, coatis, porcupines, pigs, gophers) can bury seeds or unearth buried seeds. At Mount St Helens, for example, the activity of surviving gophers in bringing soil and propagules to the surface of the sterile ash deposits facilitated plant establishment and the start of forest regeneration (Andersen and Mac-Mahon 1985). Secondary dispersal of wind-dispersed seeds of *Pinus jeffieyi, Pinus ponderosa* and *Pinus contorta* by chipmunks and other rodents in the Sierra Nevada is instrumental in moving seeds up to 69 m away from the parent tree compared with only 12 m by wind dispersal alone. Mockingbirds, doves and land iguanas have been essential in facilitating regeneration of the tree *Bursera graveolans* (Burseraceae) on the Galapagos Islands after the elimination of introduced goats. The *Bursera* seeds fail to germinate directly beneath adult trees, and recovery of the population to levels prior to the introduction of goats is entirely dependent on the native seed dispersers. Vertebrate dispersal agents may further benefit plants by dispersing mycorrhizal fungal spores in addition to seeds, as in the Pacific North-west where squirrels disperse the obligate symbiotic mycorrhizal fungi of conifer seedlings.

Dependence on animals for seed dispersal exposes the plants to the risk of dispersal failure should seed vectors become rare or extinct. The population structure of the tambalocoque tree *Sideroxylon sessiliflorum* (Sapotaceaea) on Mauritius is thought to have been affected by the extinction of the dodo, which dispersed the seeds and whose elimination doomed the tree to extinction as one of the 'living dead'. The likely cause of the dramatic decline of colonization rates of many fleshy fruited species in the remnant rain forests of the Mascarene Archipelago, in which Mauritius is located, is thought to be due to the extinction of over 50 per cent of its land vertebrates in recent decades. Cox *et al.* (1991) describe preliminary observations of pollination and seed dispersal failure for several bat-pollinated and -dispersed plants on the island of Guam where bats have suffered severe declines in recent years. Indeed, Fujita and Tuttle (1991) note that at least 289 plant species (according to current published information) depend to varying degrees on large populations of flying foxes for their propagation, and the fact that flying foxes are increasingly threatened is a cause for broad concern.

Selective logging and fragmentation of forest habitats have a large impact on vertebrate abundance and behaviour, which in turn is likely to affect seed dispersal of many animal-dispersed tree species. Forget and Sabatier (1997) reported that large frugivores avoid recently created gaps, and many bird species do not move into central regions of large gaps.

However the responses of species to fragmentation can be highly variable and the impact on forest tree regeneration is often difficult to predict. Frugivorous birds may decline sharply after fragmentation leading to decreased seed dispersal, while small mammal populations remain stable or even increase in abundance. However, the behaviour of small mammals may alter as the landscape is transformed into a primary and secondary forest mosaic. Terrestrial mammals readily move between primary and secondary forests and may promote regeneration by dispersing primary forest tree seeds into degraded secondary forest, although seeds can also be transported in the opposite direction and alter the nature of primary forest succession. Management of animal seed-dispersers may be necessary to accelerate regeneration of degraded areas and to promote the establishment of native plant species. Availability of hollows, snags, perches and nest boxes increases the abundance and diversity of birds and bats, thereby preserving dispersal functions and accelerating forest regeneration. Where vertebrates choose to sit while processing seeds is critical to the reproductive success of plants and appropriate provision of perches and nest boxes can encourage frugivorous birds to process (and defecate) consumed seeds in degraded sites. However, if tightly coupled mutualisms exist between a plant and animal, then specific management strategies may be required to ensure the continued existence of both.

Herbivory

Vertebrate herbivores are instrumental in determining the vegetational composition of forest systems. The role of the African elephant in maintaining the balance between savannah and woodland in Africa is well documented and the impact of grazing and browsing mammals on temperate forests has been reviewed by Gill (1992a,b). The loss of large herbivores from Europe in the Pleistocene is thought to have led to the transformation of short grasslands to less productive forests with concomitant extinctions of many small mammals dependent on the more nutrient-rich vegetation. More recently, the elimination of rabbits by the disease myxomatosis in Britain has resulted in many grassland areas reverting to scrub woodland.

Overgrazing by deer in European and North American woodlands causes a decline in native woodland species, facilitates the spread of invasive weeds, prevents regeneration of woody species and may result in local extirpation of herbaceous plants. Moderate grazing can lead to plant species replacements. In North America, preferential feeding of white-tailed deer on pine *Pinus strobus* results in its replacement by birch *Betula papyrifera,* which has a greater resistance to browsing.

Moderate or low levels of grazing can also result in greater vegetational diversity than either overgrazing or an absence of grazing. Thus, in tropical Mexican forests the defaunation of large mammal herbivores such as tapirs, peccari and deer by hunting and illegal trading led to increased density but decreased diversity of seedlings and saplings. Grazing represents a nutritional drain on plants and the ability to survive sustained grazing pressure may depend on soil quality and efficiency of nutrient capture.

The response of *Colophosper-mum mopane* trees to high levels of elephant herbivory in *C mopane*-dominated woodland in Zambia was related to the availability of soil nutrients; poor soil quality combined with elephant herbivory increased plant species diversity and reduced dominance by *C. mopane.* Periodic large-scale diebacks were also suggested to be a function of soil nutrient depletion from sustained grazing by elephants in these woodlands. By browsing and debarking seedlings, saplings and trees, large mammalian herbivores can influence forest plant diversity and abundance, which in turn affect the resource and habitat quality and quantity for a range of other plant and animal species. Woodland degradation in Botswana by African elephants resulted in substantial changes in bird species composition, although bird diversity was unchanged, while culling of hippopotamus in Uganda allowed regeneration of plant diversity and consequential increases in buffalo, elephant and waterbuck densities. In tropical wet forests, large mammal densities are considerably lower than in African savannah forests and vertebrate herbivores do not have a major impact on habitat structure or composition.

THE NUTRIENT CYCLE AND THE SOIL ECONOMY

The flow of water in the forest ecosystem plays the vital role of carrying the nutrients required for plant growth from the soil and controls the rate of uptake of nutrients. The nutrient cycle thus represents the chemical aspects of ecological processes of a forest ecosystem. In economic terms, the nutrient cycle describes the economy of the soil, describing and quantifying the nutrient uptake from and return.to the soil on which the forest grows.

The botanical process of plant growth requires a large number of elements like hydrogen, carbon and oxygen, macro-nutrients like Calcium, Potassium, Magnesium, Nitrogen, Sulphur and Phosphorus, and micro-nutrients like Boron, Copper, Iron, Manganese, Zinc and Molybdenum. Elements like CH and O are available from water and Carbondioxide from the atmosphere.

These nutrients with the exception of N are available from the weathering of rock mineral and Nitrogen is available from the atmosphere. Apart from the biomass exported out of the forest ecosystem, these nutrients eventually return to the soil surface through deadwood and litter fall as well as washing of foliage by rain water. On the soil surface a variety of forest floor fauna including micro-organisms and bacteria transform the biomass through decomposition and release nutrients for further plant nutrition. The uptake and return of nutrients in the forest ecosystem is well studied in the temperate regions of Europe or the USA. Unfortunately, the soil economy associated with indigenous tree species in tropical countries like India is least understood. This obviously leads to wide gaps in knowledge for the proper choice of species and their management in afforestation programmes.

In every forest ecosystem the nutrients that are used by the trees are normally returned to the soil completely. When forest biomass is extracted and transported for the satisfaction of sustenance needs or industrial/commercial demands, substantial amounts of nutrients go out of the forest ecosystem, and for intensive forest exploitation artificial fertilisation of the forest soil becomes essential.

Nutrients are supplied to the forest trees both from the atmosphere and the soil. Nitrogen is available from the atmosphere directly, as dissolved nutrients in the rain water and as particulate deposition which gets washed down to the forest floor through rainfall. Rain and wind erosion transport the nutrients from the parent rocks to the soil and the moisture in the soil dissolves them and transfers them to the body of the trees.

The nutrients are returned to the soil through the litter which contains organic remains of plants, like leaves, barks and twigs in exploited forests and organic remains of animals on the soil surface or in the top soil layer. In tropical rain forests leaf litter is about 10 tons per ha while in the open conifer forests it may be only I ton per ha. One part of green plants combine with the litter as animal waste through the consumption of green matter by the herbivores.

The accumulated leaf fall and other forms of litter then begin to decompose on the forest floor through the action of micro-organisms present in the soil. In tropical conditions where soil biotic activity is encouraged by relatively higher temperatures, the rate of decomposition is quite rapid. Due to the content of the leaves and the soil chemistry, the rate of decomposition of litter in rapidly growing tropical forests is several times greater than in the conifer forests in the temperate regions. The soil organisms that decompose the litter are mainly bacteria and they multiply in soil with earthworms. The soft parts of the plant are normally decomposed by micro-organisms alone but woody biomass are broken. down by a complex interaction, thus the return of the nutrients back to the soil is an intricate process involving many actors. As decomposition proceeds, the nutrients are released in the form of soluble ions that can be directly absorbed by the root system and the cycle starts once again.

The nutrient cycle is disturbed by the destabilisationof the hydrological cycle. With the opening up of the forest canopy and instant surface run-off increasing, the leaching of the nutrients increases and the nutrients available for new plant growth become less, thus setting in motion a process of decay in the forest ecosystem. In extreme cases of nutrient loss and continued exploitation of the forest biomass, the vegetational evolution is reversed and a full canopy forest gets degraded to scrub forest or grasslands.

Both the water economy and nutrient economy constitute nature's economy in forest ecosystems. They need to be stable in order to sustain the productivity of forest ecosystems. The two other economies, *i.e.*, the survival economy of the basic needs satisfaction of the people and the market economy of forest product demand of the industrial/commercial sector compete for the same resource base, forest biomass and generate conflicts over forest resources between the needs of nature and the people on the one hand; and between the needs of nature as well as people and market demand, on the other.

In the context of the forests of India, nature's economy and survival economy have always been overlapping and were simultaneously functioning without major conflicts as the small survival needs of the people were satisfied through a conservation-oriented utilisation managed by an informal but strict code of conduct towards forests. It is thus reasonable to assume in the context of forest resource utilisation in the precolonial periods, that the satisfaction of survival needs was an intrinsic part of the functioning of forest ecosystems. This was particularly so because human settlements in India grew as an integral part of the forest ecosystem and not at the cost of it as was the case in industrialized countries in the last few centuries.

With the introduction of large-scale commercial exploitation of forests by the British this situation underwent a drastic change. A schematic picture of the three competing biomass requirements of the three economies. The horizontal axis represents the distance (D) from the core of the forest ecosystem while the vertical axis represents the quantity (Q) of biomass required by the three competing economies. Nature's requirement (ON) is spread throughout the forest ecosystem while the survival requirement (QS) is divided between inside and outside the forest ecosystem. It should be noted that the spread is not too far away 'from the forest ecosystem's boundary since only the local population can collect the' forest biomass. The requirement of the market economy (QM) is high as well as spread over long distances far away from the forest ecosystem, since it can be transported over long distances. This indicates a continuous long distance transfer of large quantities of forest biomass outside the forest ecosystem. All forest related conflicts are thus based on conflicts between the above mentioned requirements-ON, QS, and QM. The objectives of forest management decide as to what should be the actual biomass quantities allocated to these diverse requirements.

When forests are viewed as a complete ecosystem and not as a mechanical collection of wood producing trees, the management strategy of forests has to evolve along the ecosystems concept. The ecosystems approach has the objective of ensuring sustainable production of an optimum biomass mix so as to satisfy the demands of nature's economy and survival needs and produce biomass for commercial/industrial purposes to the extent possible. Accordingly, such an approach should

- Lead to a harmonious utilisation of resources not resulting in external diseconomies (exploitation of forests for timber should not lead to floods and silting).
- Provide for the equitable returns to various sections of society depending on the same resource system (exploitation of forests for timber should not conflict with the domestic requirements of fodder or wood as long as alternatives are not made available at a low cost).
- Wherever externalities cannot be eliminated, the logic of costs and returns needs to be worked out for proper methods of compensation to those who bear the costs generated by activities that benefit others.

EFFECTIVE ECOLOGICAL RESTORATION

Ecological restoration for protected areas will, and indeed should, be motivated primarily by the desire or need to restore natural (and associated cultural) values of the protected area related to ecosystem structure and function (*i.e.*, ecological integrity), as the objectives of restoration draw on the original purposes of a protected area and objectives for management that are often described in management plans. Associated cultural values (*e.g.*, cultural heritage values such as historic buildings or viewscapes as well recreational, aesthetic or spiritual values) may be restored simultaneously. Regardless of the specific values being restored, achievement of ecological restoration goals also requires attention to the underlying causes of degradation, opportunities for restoration associated with human knowledge and cultural practices interactions with wider landscapes and seascapes, and careful monitoring to learn from experience and facilitate adaptive management. These concepts are associated with the principle of effectiveness.

Effective ecological restoration for protected areas is restoration that re-establishes and maintains the desired values of a protected area.

EFFICIENT ECOLOGICAL RESTORATION

Ecological restoration can be a complex and relatively costly process in terms of time, resources and effort. However, it can also help to achieve significant benefits both inside and outside protected areas related to, for example, climate change adaptation and mitigation and socio-economic and cultural well-being. Evidence suggests (Nellemann and Corcoran, 2010) that well-planned, appropriate restoration can have very high (up to 75:3) benefit:cost ratios in terms of return on investment.

Ecological restoration for protected areas should thus aim to maximize beneficial outcomes related to ecological, social-economic and cultural dimensions of the system and minimize costs, while not losing sight of conservation goals. This may involve prioritizing restoration efforts according to various locally-determined criteria. These concepts are related to the principle of efficiency. *Efficient* ecological restoration for protected areas is restoration that maximizes beneficial outcomes while minimizing costs in time, resources and effort.

Engaging Ecological Restoration

Protected area managers can play a vital role in implementing ecological restoration projects and programmes that can help to address the immense challenges posed by large scale ecological and social change. But they cannot do this alone. Collaboration among community stakeholders is key to the long-term success of ecological restoration efforts at every scale. Engaging and involving stakeholders and protected area visitors in restoration planning and activities is important for creating a constituency of support for restoration projects, reconnecting people with nature, and building partnerships to reduce causes of degradation in the larger ecosystem. Collaboration with communities and other agencies is the foundation of any initiative involving restoration beyond the borders of a protected area, for example to improve connectivity or address watershed scale issues.

In situations where protected areas have resident or local communities, indigenous peoples, or other stakeholders, ecological restoration can only be successful in the long-term if supported by these groups. In some countries, there is a legal obligation to consult the public about projects that impact natural resource use; guidelines are also available. In all cases it is best practice to obtain free, prior and informed consent from traditional and indigenous peoples for projects on their traditional lands; many community-run protected areas also involve interactions between different stakeholders. Regardless, public engagement, reciprocal learning and listening can create a constituency of support and provide the stable long-term foundation for success.Local and traditional ecological knowledge brings valuable additional practice, information and understanding to restoration activities. Dialogue can also help the design of restoration projects to maximize community benefits, thus reinforcing support for restoration and for protected areas.

Outreach, consultative and collaborative processes raise awareness of potential problems and can help engage communities and other stakeholders in restoration and long-term monitoring. Restoration is a good tool by which to engage stakeholders in nature, to create awareness and to provide people with hope that degraded areas are not permanently lost.

Engaging people in restoration efforts in protected areas is an opportunity to reconnect people with nature and provide them with a spark of hope to start restoration in their own communities. At its best, the process of restoring an ecosystem can bring people together, often in significant ways that lead to a renewed engagement between society and ecological processes. These concepts are related to the principle of engagement. *Engaging* ecological restoration for protected areas is restoration that seeks opportunities to collaborate with surrounding communities, to promote stakeholder participation and to enhance visitor experience.

ECOLOGICAL RESTORATION FOR PROTECTED AREAS

This publication is aimed principally at protected area managers at both system and site level. It provides guidance on how to deal with diverse challenges associated with restoring ecological, cultural and other important values of protected areas. Restoration activities beyond the border of a protected area are also in some situations beneficial and even critical to restoring the protected area values; the document thus uses the term 'restoration for protected areas' to refer to restoration activities conducted both within and around protected areas.

This guidance document does not attempt to be a comprehensive restoration manual and, while some elements assume a relatively high degree of technical understanding, it does not provide detailed instructions on specific methodologies and techniques. The document starts with an introduction to the key *concepts* that underpin the fields of restoration and protected areas management and that are relevant to conducting ecological restoration in protected areas.

Underlying *principles* for restoration that are intended to govern the work of individuals and organizations responsible for protected area management, along with their stakeholders, in setting restoration policies, goals, and objectives and in guiding their actions regardless of biome-, site- or issue-specific circumstances. The principles provide the foundations for *guidelines* that provide advice in general terms on how ecological restoration for protected areas should be approached such that it remains consistent with the underlying principles, while allowing for biome-, site- or issue-specific variation in implementation.

This guidance document then draws upon the experience and lessons learned in the restoration of ecosystems generally, and protected areas in particular, to identify *best practice* methods and techniques for planning and implementing restoration projects. Drawing on these principles, guidelines and best practices a seven-step framework that identifies recommended decision-making *processes*for carrying out ecological restoration for protected areasis elaborated.

RESTORATION PROCESSES FOR PROTECTED AREAS

The principles, guidelines and best practices outlined in the preceding chapters provide an overview of appropriate approaches and methods protected area managers and their partners should use in implementing ecological restoration projects and programmes in and around protected areas. We summarize how those approaches and methods can be brought together in a planning and implementation framework that will help to ensure restoration success.

Ecological restoration, like any management action, should be carried out within the context of the overall management of the protected area, the protected area network, and the surrounding landscape or seascape. A number of factors can influence decisions such as whether restoration is an appropriate intervention, whether it is a relatively high or low priority for the specific site or system, who should be involved, and what appropriate goals might be.

An evaluation of information such as management objectives for the site and relevant local or national policies and legislation isan obvious starting point. A review should be made of any regional and international conservation strategies, goals, programmes and policies which could help define or impact the restoration project. For example, national, regional or global action plans associated with issues such as invasive species or climate change adaptation and mitigation may influence the selection of restoration goals. Key natural values of a protected area are generally listed in planning documents (ideally management plans) or information from the time of designation or establishment.

The values listed will tend to be the management *priorities* rather than all natural values; in many cases addressing a target, such as an endangered species, will require wider ecological restoration. Many protected areas also have cultural heritage values such as sacred natural sites or historical remains which should be respected in restoration. In some cases, natural and cultural values of the protected area may be intertwined and can benefit simultaneously from restoration actions. For example, healthy forests, streams, and salmon populations are important natural values that are being restored in Gwaii Haanas National Park Reserve of Canada. They are also important cultural values for the Indigenous Haida Nation in whose territory the park reserve is located.

Regardless of the management priorities or objectives for the protected area, preliminary information about the site or nearby sites, including, for example, results of similar restoration efforts elsewhere, the attitudes of local communities, the degree of interest and support of potential partners of resources for long-term monitoring of results should also be evaluated to determine whether restoration is feasible and appropriate.

In the absence of formal, written management objectives, information on the overall focus of management could influence the approach to restoration. If the IUCN protected area management category has been identified for the site, this provides valuable information on the overall priorities of management. The IUCN category is selected primarily by management objective; that is, it relates to the aims of management rather than the current status, so that protected areas in any category can be subject to restoration. For example, loss of ecological integrity may be a trigger for restoration in a protected area managed according to category II objectives whereas threats to a natural feature may trigger restoration in an IUCN category III site.

Threats to interactions between culture and nature (*e.g.* survival of traditional, nature friendly management systems) or to sustainable uses (*e.g.* fisheries) would identify the need to restore in protected natural areas that are managed for values related to natural landscapes or seascapes or to the sustainable flow of products and services. In practice the category usually also implies something about the protected area status; some types of long-term, active restoration are unlikely to be suitable for every category. For example, an area requiring *indefinite* active management interventions to maintain wilderness values would not generally be categorized as a wilderness area.

Where restoration of such areas is identified as a priority it should be approached with particular caution, both because of the risk of causing more damage to a relatively intact system through the restoration process, and because of perceptions about the implied status of the protected area.

Regardless of the broader protected area management context for ecological restoration, the process of planning and implementing an ecological restoration project should be iterative. Presented below is a compilation of best practice processes drawn from restoration projects and experiences around the world. The framework outlined here puts considerable emphasis on the planning and design elements of developing effective, efficient and engaging ecological restoration. This emphasis is particularly important for projects in areas where restoration has not been previously carried out or where there is little practical experience in successful restoration. Above all, however, stresses that adaptive management is a vital component of ecological restoration.

All protected area management should be flexible enough to adapt if circumstances change – if new information comes light or management interventions do not have the expected results. Restoration projects need to be more flexible than most, as our knowledge remains incomplete about how ecosystems work, how their components interact, and how human interventions can and should influence relevant processes. This incomplete understanding is further complicated by the impacts of climate change, which mean that much of the knowledge we have gained about how natural systems work and the way they interact with the socio-economic systems in which they are embedded is becoming less certain.Restoration projects can thus be seen as experiments that help us learn about how ecosystems are put together and how they function so that we can continue to improve our understanding of how we should intervene. Any ecological restoration project are sequenced along the first loop of the helix and represent the first full cycle towards a restored ecosystem. The three core principles of restoration – effective, efficient and engaging – are tied together to show the importance of an integrated approach. Subsequent loops reinforce successful restoration projects as those that regularly to ensure original goals are being met (or goals need adjustment) and engage effective monitoring and communications. In time, the loops in the helix become smaller to represent decreased effort, money, energy and data to sustain a restored ecosystem.

DEFINE THE PROBLEM AND ENGAGE STAKEHOLDERS

Define the Restoration Problem that Needs to be Addressed

Building on the information available a problem statement should be developed as the starting point for identifying why restoration is an appropriate and effective response to an issue. Where routine monitoring is in place, a restoration problem may be identified when a particular measure of the natural or other identified value of the system falls below a pre-determined threshold. In other cases, anecdotal information may clearly indicate that protected area values are degraded or are becoming degraded and that restoration would help to reverse this process. Regardless of the information base available, the problem statement should consider:

1. The problem: a description of the problem (*e.g.*, decline in target wildlife, emergence of invasive species, loss of key habitat, loss of cultural landscape), including where possible quantitative data on the scale of the problem, timescale and rate of change.
2. The causes: where known, a description of the immediate and any underlying causes of the problem (*e.g.*, immediate cause of wildlife decline might be illegal poaching, underlying causes might be poverty, lack of support for the protected area management or a thriving bushmeat trade).
3. Why restoration: a justification of why restoration is appropriate and likely to be successful. It should provide evidence that the restored habitat or species will not succumb to the pressures that caused the original problems. There is, for instance, little point in spending resources to increase wildlife numbers if poaching remains unchecked.

The problem statement is an important component of adaptive management. If defined precisely enough, the problem statement will also help to define the necessary responses and the monitoring requirements needed to assess progress.

Ensure Effective Stakeholder Engagement

The identification and engagement of relevant stakeholders and partners (*e.g.* local communities and indigenous peoples, government agencies, universities, researchers, landowners, conservation groups, tourism bodies, recreation specialists, local experts, and the general public) is critical to successful restoration. The third principle of ecological restoration and associated best practices provide guidance on engaging stakeholders in restoration projects. Meaningful engagement may range from informing, through involving/consulting, to collaborating.

Early consultation is fundamental to the success of ecological restoration projects, not only in conducting effective project planning but also in building a greater understanding and appreciation of the role of ecological restoration in attaining broader conservation goals. Engagement therefore should start with involvement in the development of the problem statement and may be further supported through early development and implementation of appropriate communication strategies.

It is important to identify the full range of partners and stakeholders who should be engaged, while recognizing that the degree of engagement will vary among groups (*e.g.* indigenous people, local communities, government agencies, universities, landowners, local and regional businesses, NGOs, tourism bodies, visitor experience specialists, local experts, and the public). Particular attention should be paid to including those with knowledge of local ecosystems and the driving forces behind degradation, and to empowering and including disenfranchised resource users (*e.g.* women and the elderly), the displaced and powerless.

Stakeholders and partners should be engaged in affirming the natural and cultural values of the ecosystem within the context of the protected area and its natural region. In particular, projects should ensure the free and prior informed consent from owners and stewards of land and water proposed for restoration. Consultation in later stages of the project should involvesharing information about the ecosystem, setting project goals, defining objectives, gaining permission for the proposed work, and contributing skills, knowledge, financial, and human resources to the development, implementation, maintenance, and monitoring of the project.

An important starting place in stakeholder engagement is to ensure that there is information and knowledge on the local, regional and national social and cultural contexts where restoration takes place. This would include history, conflicts and resolutions, organizational networks, stakeholder relationships, institutional priorities and what has (and has not) worked in the past for conservation and restoration.

Projects should consider a broad range of tools to engage partners and tailor these to the experience of the stakeholders and complexity of issues (*e.g.*, workshops, village meetings, open houses, events, modelling and scenario workshops, temporary programme office, advisory committees, use of local or regional planning processes; study exchanges; etc). The cultural environment of a country will determine the range of engagement tools. There should also be awareness that relationships can be delicate and sensitive and some skill in negotiation may be needed. Transparent and pro-active conflict resolution processes should be established among the parties involved in the restoration process. For example, agreement should be reached at the start of the project on a conflict resolution processes and either a person should be nominated who will mediate a way through any conflict or there should be an agreement on the way in which a mediator would be selected (Australian Heritage Commission, 2003).

Finally, of course, it should be remembered that effective engagement needs to be efficient as the resources of the project and the stakeholders themselves can easily be over-stretched if engagement processes are not well planned and executed.

Develop a Communication Strategy

Ecological restoration interventions should be supported with active communication and outreach programmes focused on the initial causes and pressures leading to the degradation, the effects of degradation, and the benefits of restoration. This should target local communities and stakeholders, visitors, and also the protected area's employees, who have contact with the public. Early and strategic communication about sensitive issues such as the reduction of hyperabundant populations of species can be particularly important in gaining public support for subsequent actions.

Communication strategies should identify type of information to be shared, how often and with whom, including the various purposes of communication (*e.g.*, engaging the public and neighbours, information sharing, routine reporting); communication methods (*e.g.*, the media, interpretativesigns, community-based special events, web sites and publications, refereed literature, and presentations at conferences and meetings); the target audiences and the frequency of communication. This should involve developing strategies for appropriate stakeholder engagement.

Assess the Problem

To be able to develop clear goals, objectives and action for ecological restoration problems it is first necessary to carry out a more detailed assessment of the restoration problem; by reviewing site conditions, developing an understanding of what the site condition could or should ideally be, and assessing the potential impacts, both environmental and social, of restoration activities. This understanding, including:

- Assessing condition
- Identifying reference ecosystem(s)
- Carrying out environmental impact assessment if necessary
- Developing data management systems

Assess Condition

Once the problem statement has been developed, but before detailed planning can proceed, preliminary information about the state of the ecosystem should be collected to assess its condition. Additional information may include projected climate trends and impacts and relevant social, cultural, economic and political data (*e.g.* relevant traditional ecological knowledge; relationships between local communities and the protected area; predicted demographic changes; economic importance of the site; political support and related governance issues). Ideally, there should be enough information to understand the extent to which pressures are changing the ecosystem; that is to determine the degree to which natural values or other important values of the system differ from those of a suitable.

In some cases, existing monitoring and assessment frameworks may have been in place for a long enough to detect damage to ecosystem structure and function or to other natural or cultural values of the protected area. The identifies ecological integrity indicators (modified from Parks Canada, 2008) which represent what might be considered an *ideal*information base.

In practice however many, protected areas will have to make decisions based on far less data. Additional information about the protected area ecosystem and its regional and landscape context may be drawn from a variety of sources including data from similar ecosystems.

Identify Reference Ecosystem(s)

In assessing and defining the problem, particularly in heavily degraded or altered ecosystems, is finding and agreeing a reference ecosystem to act as a comparison and "ideal" against which to restore. These will often be undisturbed sites in similar ecosystems, descriptions of such sites, or documentation describing the target state of the restored ecosystem. For example, in the case of restoration of Gwaii Haanas National Park Reserve an old-growth forest where no logging has occurred serves as the reference ecosystem. If restoration aims to bring back a cultural landscape or seascape, the reference may be a similar cultural ecosystem. Outside of large, homogeneous ecosystems, reference ecosystems are seldom if ever exact replicas of what restoration might achieve, but rather provide a broad picture of likely ecosystems. Given this complexity, project managers might consider identifying and describing multiple reference ecosystems. In such cases, objectives and targets would be described with a range of possible outcomes in mind, recognizing the inherent variability of natural systems and that unforeseen or uncontrollable disturbance may have an impact on the outcome.

Finally, many degraded ecosystems can no longer feasibly be restored to a pre-disturbance condition, particularly under climate change. As has been discussed elsewhere in this document, a more realistic and desirable aim here may be to use an understanding of the historic ecosystem as a guide to recovering a resilient ecosystem with structural and functional properties that will enable it to persist into the future. The degree to which historical, present or future conditions are described in reference ecosystems will depend on the management objectives for the protected area, as discussed above, and the goals and objectives of the restoration project, as discussed below.It will also depend on the degree to which the protected area is experiencing, or is predicted to experience, rapid change related to climate change and/or other stressors.

CARRY OUT AN ENVIRONMENTAL AND SOCIAL IMPACT ASSESSMENT, IF NECESSARY

The planning of an ecological restoration project must also include consideration of the potential for adverse environmental effects, for example from alteration of ecosystem structure or function, introduction of infrastructure, or human presence during the restoration. An environmental impact assessment will aim to identify all the potential consequences of a project, unintended as well as intended, in order to maximize benefits and minimize adverse effects. This is an element of good planning, regardless of whether or not it is required by legislation or policy, although if there is a legal obligation the assessment should be done in a way that serves the needs of that process. In general, a good environmental impact assessment is one that supplies useful information to planners and decision-makers. It need not be lengthy: environmental impact checklists are available and may be suitable. Advice should be sought from an assessment specialist early in concept development on how and when to conduct an efficient and useful impact assessment and who should be involved.

As well as understanding the environmental impacts of a restoration project it is important to recognise that ecological restoration projects have potential social and cultural impacts, both positive and negative, that need to be identified and addressed early in the planning process. Ideally, ecological restoration projects should contribute to sustainable. On the positive side, restoration can bring back ecosystem services, sustainable supplies of natural resources, aesthetic qualities and increased benefits from ecotourism. But it may bring costs, such as unwanted controls on natural resource use, or inadvertent damage to socially and culturally important sites. Identifying potential costs and benefits early in the process can avoid later problems.

Particularly in protected areas that have indigenous populations it is vital to consult and follow applicable national constitutional, legislative and international obligations to determine the duties and principles for engaging indigenous communities and their governments. When assessing the restoration problem it is crucial to understand all people's views and take their degree of dependence on the ecosystem into account. This includes the socio-economic and cultural links between people, species and landscape priority areas and the utilisation of 'resources'. One of the ultimate goals of restoration can be the re-establishment of traditional cultural values and practices that contribute to the ecological, social and cultural sustainability of the protected area and its surroundings but this will only be effective if the cultural and social dynamics of the site are understood before project planning begins.

Develop Data Management Systems

The gathering of considerable background information (*e.g.* research papers, policy documents etc). Data management, whether digital or analogue, is essential to understanding, planning, implementing and adapting ecological restoration projects. In addition, collection and archiving of data is critical to ensuring the success of future projects. Plans for managing data should be included early on in the project. Ideally sites will already effective archiving systems which the restoration project can add to. If not, important factors to consider in the context of ecological restoration projects, which in many cases may span years or decades, include:

- Accepted metadata standards, using a records management (archival) system to identify data/records locations and ensure their effective retrieval
- Ensuring data/records are secure regarding access restrictions, intellectual property rights, and use of data sharing agreements where applicable (note this is of particular relevance where TEK is shared and employed)
- Clearly defined and rationalized data analyses that are specific about biases in collection and analysis and limitations
- Developing reference collections where digital photographs are taken, vouchers are collected (if possible) and identifications of taxa are peer reviewed in some form
- Data management plans should also address data integrity, digital file maintenance, and data migration and include plans for efficient data and information sharing within and among protected areas agencies.

DEVELOP GOALS

Develop Restoration Goals

One of the most important steps in the ecological restoration process is the development of effective goals that will guide project planning and implementation. Protected area managers should work closely with stakeholders to develop restoration goals that are consistent with the IUCN WCPA Ecological Restoration Principles and Guidelines such that they provide the foundations for measuring effectiveness, efficiency and engagement.

Ecological restoration goals should be clearly stated, realistic and achievable, based on a shared vision of the future state of the ecosystem, between protected area managers, stakeholders and partners. The preparatory help project managers reach the stage in the process where goals can be realistically developed. All principles, guidelines and best practices of ecological restoration, should guide the goal-setting process along with the information collected above.

For example, the principle of effectiveness will help to guide the selection of goals related to the recovery of specific values of the protected area.The principle of efficiency, and the associated guidelines and best practices related to capacity, should help define the barriers within which the project needs to operate, and hence which possible goals are realistic.The principle of engagement should remind managers that for restoration to be successful over the long-term, goals related to community understanding, support and appreciation of the protected area may be equally important as those associated with the recovery of specific natural values of the system. Project goals should be realistic and achievable in the context of off-site influences, broader ecosystem functioning and global change. For example, because of the often-significant mobility of many marine, large mammal and bird species, their restoration may be beyond the ability of protected area managers alone, and require collaboration and coordination with other resource managers.Similar concerns exist regarding projects aimed at assisting the recovery of migratory species or in restoring freshwater ecosystems that are affected by what happens elsewhere in the watershed. Where collaboration is fundamental to project success, it should be identified as part of the project's goal. For example, Habitat 141°: Restoring habitats and linking protected areas in Southern Australia specifically aims to work collaboratively to restore and connect the wider landscape and enhance the natural and cultural values of existing protected areas in Southern Australia.

Goals for individual restoration projects should be linked to, and consistent with, all relevant national, regional and local policy and management planning goals and should reflect global goals and policies.

For example, with a goal of restoring forest habitat and natural ecological processes and implementing carbon sequestration projects to mitigate climate change, restoration of protected areas in the Atlantic Forest of Brazil is contributing to local, regional and international biodiversity and climate change-related policies and goals.

While one community or organisation may initiate the project, it may serve the needs of multiple communities/organisations. Linkages are particularly important where there are complexes of natural areas and other types of green or open space, held under different ownerships, that all contribute to the ecological integrity of a larger landscape (*e.g.* biosphere reserves or transboundary protected areas). Establishing these linkages early will create efficiencies and ensure the project is compatible with large-scale plans and processes. In many cases, multiple competing goals for the restoration of protected areas ecosystems may exist. For example, goals for the recovery of sea otter populations may conflict with goals for the sustainable harvest of shellfish in marine protected areas. Potential conflicts should be considered and resolved as goals are established.

DEVELOP ECOLOGICAL RESTORATION OBJECTIVES

An overall picture of what restoration is trying to achieve; the *objectives* should provide the details about the individual actions needed to achieve the goals. Objectives should where appropriate consider both ecological and cultural outcomes. They should be specific enough to be measurable through monitoring. For example objectives may be that: primary productivity meets a specified level; a specific percentage of an invasive species is removed; species population size is within 95 per cent confidence limits of reference conditions etc. They should also be achievable within an acceptable range of variation, and consistent with other relevant protected area plans, policies and legislation. If it proves impossible to develop objectives that meet these criteria, it may be necessary to think again about the problem definition and project goals.

Restoration within a protected area can have a range of objectives. In some cases, objectives may pertain principally to the restoration of natural ecosystems. For example, the Lintulahdet Life Project – restoring wetlands in Finland had objectives related to the re-establishment of open water areas, removal of invasive species, creation of insect habitats and removal of small non-native mammals to improve the breeding success of wetland birds.

Similarly the objectives for the restoration of Asiatic Black Bear in national parks in the Republic of Korea were primarily ecological. In contrast, the Fandriana Marolambo Forest Landscape Restoration project in Madagascar was set up to restore and protect degraded forest and address community pressures that were leading to degradation. Project objectives were related to conserving the unique biodiversity and functional integrity of the forest, restoring forest goods and ecological services, and improving the well-being of the people living in the area. Objectives for the restoration of protected areas in the Atlantic Forest of Brazil are similarly diverse.

The development of objectives should be based on a sound understanding of the condition of the area intended for restoration and actions should be based on best practice for the specific restoration goal.

The complexity of a project will affect the number and type of objectives required. A complex project is more likely to require specific objectives for societal engagement. Relatively simple projects may have only a single goal and a few objectives. If there are multiple linked objectives, their relationship should be described along with the order in which they are to be pursued, and whether or not they can be pursued concurrently. The development of a conceptual model may help to organize and focus the planning process and assist in the development of specific objectives and testable hypotheses.

Develop Restoration Approach

Specific actions (*i.e.* restoration activities) to achieve individual objectives are developed in the restoration process. Developing the approach includes:

- Defining the scope of the project
- Developing an implementation plan
- Develop monitoring plans

Define Scope

The scope of the project should be defined in terms of both the geographic area involved and the timescale, in consultation with stakeholders and partners. In some many cases, restoration projects conducted primarily inside a protected area (*e.g.*, invasive species removal) will also be dependent on actions outside the protected area (*e.g.*, transportation of invasive species into the region). The scope of such projects, particularly as they pertain to engagement, thus extends beyond protected area boundaries.

Consideration of timescale is critical in that some objectives (*e.g.* species reintroduction) may be quickly attainable whereas others (*e.g.*, reforestation) may take decades. While most restoration efforts are necessarily long-term, the period of active intervention in restoration differs markedly among biomes. Fast-growing, tropical vegetation and fast-breeding animal species are generally easier to re-establish than slow growing vegetation or animals that breed slowly, such as some birds of prey. Objectives to be included in the project scope should be attainable with the resources available.

Timescale is also important to consider in the context of community engagement in that restoration may take place over several generations and plans may need to be made for considering how to ensure engagement is of the right intensity and duration to support the project over the long-term.

Goals and objectives established are used to define broadly the interventions required. There are usually several options available for achieving these objectives. These options should be considered here, along with relative costs and likelihood of success. For example, a project could be designed to achieve the maximum potential restoration as quickly as possible or designed to achieve restoration over a longer time, with or without assistance. A short-term intervention, such as stabilization of a rapidly eroding site, may be needed while longer term plans are developed. Decisions may be helped by a risk assessment approach. Potential risks (*e.g.*, risk of failure, permanent loss of a resource, cascading effects, off-site impacts, impacts on visitor experience, or losing support of partners) should be evaluated and gaps in knowledge which may lead impact the project should be identified.

Develop an Implementation Plan

The development of a detailed plan for implementing a restoration project can vary from simple to complex, depending on the aims of the project. For restoration projects with an emphasis on research and lesson-learning, at this stage a conceptual model may need to be developed, with associated hypotheses, against which to test and measure progress. For those restoration projects with a less academic focus, this stage just involves identifying the broad stages needed to undertake the restoration and the associated best practice steps to be followed, drawing on these guidelines and technical guidance referred to in the text and resources.

Regardless of complexity, designs for ecological restoration projects should be consistent with the "learning by doing" approach of adaptive management. The responses of ecosystems to planned restoration activities cannot be predicted with certainty, therefore restoration embraces the concept of adaptive management, in which, ideally, restoration projects are implemented as deliberately conceived experiments.

Effective adaptive management requires:

- The setting of time-bound targets for interim and final outcomes;
- Monitoring of performance measures to track progress; and
- The setting of intermediate thresholds for consideration of success or the need to change actions or policies.

Decisions regarding appropriate management strategies, or changes to such strategies, should be made on the basis of measured results. Reference ecosystems and conceptual models should be useful in establishing these targets, measures, and thresholds. The extent to which individual projects have the time or energy to adopt such a research-orientated approach will vary of course, but the key concept of learning by doing is important to retain.

Drawing as appropriate on elements of conceptual and practical design described above, detailed restoration plans should be developed to identify and list the actions needed to achieve restoration, covering both the selection of specific restoration treatments and the tactics and technologies used to implement them. Guidelines for ecological restoration should be referred to in order to ensure proposed treatments are consistent with recommended approaches.

This should include a work plan detailing roles and responsibilities, decision-making authority, onsite supervision and workforce, logistics, permits and safety concerns. Locations of the work should be specified along with the timing and costs of each activity. Plans and budgets have to consider contingencies (*e.g.*, weather, availability of nursery plants or other biotic resources) wherever possible. Plans for implementation monitoring should be included. The planning for the involvement of stakeholders and partners in implementing the restoration project will contribute to its success.

Develop Monitoring Plans

Monitoring of restoration will often be linked to other monitoring in protected areas, and should consider other work in the protected area or surroundings to identify possible overlaps, optimize monitoring programme design and resource expenditures, and contribute to reporting. Existing monitoring systems may be able to provide information on whether or not restoration is working: for instance monitoring of water birds could tell a lot about success of wetland restoration. Ideally, monitoring, evaluation and adaptive management should be conducted at a scale that is appropriate to capture ecosystem-level characteristics. Local stakeholders, including indigenous peoples, can sometimes be the most effective collectors of monitoring data, either as part of collaborative agreements or for a fee. However, the focus should be on measures and strategies that are specific to restoration and its desired outcome. It is important to agree and record precise details on when and how monitoring is carried out.

Given the long time period of restoration projects it is highly likely that different staff members will be involved in monitoring; it is therefore very important to ensure monitoring continues identically, otherwise results are likely to differ with recorders.

Monitoring should be directly integrated into the design of the restoration project, ensuring that all stakeholders understand and agree with indicators, which reflect their concerns (Estrella and Gaventa, 1998). Monitoring should include indicators/measures that are:

- Related to objectives;
- Accurately measurable;
- Appropriate to the temporal and spatial scale of the ecosystem; and
- Cost-effective.

For effective adaptive management it is also important to evaluate progress being made towards interim targets. This enables decisions to be made on whether to continue with the current approach, to adapt it or to formulate additional research questions. Interim reports may also be important as a means of demonstrating results to retain community, political or financial support and strategies disseminating the results of monitoring should be included in communication plans. The strategies for monitoring apply equally to ecological and socio-cultural objectives. Predicted expenditures should also be monitored and budgets re-evaluated during the project. The restoration design should also consider how and when detailed monitoring of the project should be phased out or incorporated with other on-going monitoring.

Although monitoring should whenever possible be constant over time, it is also important to review indicators on a regular basis to ensure that they are still up to date. Ideally all research methodologies and data should be freely available.

Existing monitoring protocols and manuals should be consulted in selecting performance measures, determining monitoring frequency, level of detail and duration, and evaluating relative costs.

IMPLEMENT RESTORATION APPROACH

Carry Out Restoration

Effective planning of an ecological restoration project following the process should facilitate effective, efficient, engaging implementation, with genuine engagement of partners and stakeholders, local communities and the general public. The restoration plan is implemented. Monitoring of measures identified conducted to assess restoration success using an adaptive management approach and modifications to the plan are made as necessary.

Communicate Progress as Per the Communication Strategy

Communication with stakeholders and partners should continue throughout the project duration,using strategies. Both successes and failures need to be reported, to encourage learning and refinement of restoration techniques and processes. The need to communicate results underscores the value of using an adaptive management approach in which progress towards meeting objectives is evaluated at intermediate stages. Communicating the achievement of short-term objectives and goals rather than waiting until longer-term objectives are met, is important to maintaining enthusiasm and ensuring ongoing engagement of partners and stakeholders.

Communication to the public contributes to a broader understanding of the concept of ecological restoration and builds public support. Communication amongst restoration practitioners helps build up the larger body of knowledge that leads to advances in this field and the development of evidence-based conservation. Communicating results to policy-makers and decision makers helps ensure ongoing support and funding and is particularly important in ensuring the long-term funding needs of complex projects that may require ongoing maintenance and intervention are met.

Adaptive Management

The effort required to design and execute monitoring programmes, to collect, evaluate, analyze, interpret, and synthesize data, and to report results should not be under-estimated. The are include:

- Continuously monitoring
- Evaluating monitoring results
- Adjusting as necessary
- Communicating results and continuing the process of stakeholder engagement

7

Community Forestry

About 43 per cent of the energy consumed in the Third World is being met from wood. In India, biomass constituted 85 per cent of the rural energy and with a per capita consumption of 1.0 ton/year, about 50 per cent of it was collected from forests. Apart from supplying wood, forests have also helped in conserving biodiversity, agricultural productivity and a safe environment.

Historically, forests provided substantial support to the rural economy. Over 60 per cent of the Indian population was directly dependent on forests for fuel, fodder, fibre, timber and a wide range of food and medicinal herbs. The forests provided year round employment to 20 million people through collection of Non-Wood Forest Products (NWFP). For over 50-60 million people representing 250 tribal communities, forests formed a part of their culture and a natural way of life.

In India, out of the total land area of 329 million ha, 143 million ha were under agriculture and 77 million ha were classified as forests. The country had a distinct and varied vegetation from the tropical wet evergreen forests of Andaman and Nicobar islands to the dry Alpine forests of Himalayas. However, with only 2 per cent of the world's forests, there was severe pressure in serving 15 per cent of the world's population. The woodfuel consumption during 1970-80 was around 175 million tons/annum and most of it came from forests and village common lands. Apart from fuel, forests provided 45-60 million m^3 timber and 45-60 million tons of forage annually. It was the traditional wisdom and joint responsibility of the communities that enforced necessary rules and regulations on the local people for sustainable management of their forests. However, with increasing growth of human and livestock populations and a shift in the ownership of natural forests from the princely states to the Federal Government, the control and moral pressure on the local population was relaxed, leading to indiscriminate abuse of forests since the middle of this century.

Unable to meet their basic needs from agriculture, many unemployed and poor families turned to forests not only for fodder and fuel, but also to generate cash income through sale of wood and other forest products. Vested interests also took this opportunity to exploit forests for commercial purposes.

Apart from biotic pressure, the following factors contributed to the denudation of forest resources (GOI, 1984).

- Increase in human and livestock populations;
- Poor management of forest soils;
- Inadequate scientific and technical inputs;
- Inadequate skills and training of the staff to play their expected new roles;
- Poor investment on forest development;
- Damage caused by mining, irrigation projects, industries, roads and shifting cultivation;

The abuse of forests continued unchecked till mid 1970s, when the ill-effects of deforestation were prominent in the form of fodder and woodfuel scarcity, soil erosion, flash floods, water scarcity, loss of precious flora and fauna and climate change. By 1981, the area under wastelands in India was estimated at 93.70 million ha, excluding the degraded agricultural lands measuring over 85 million ha. Most of the natural forests around the villages had turned into wastelands reducing the area under forests to only 40 million ha. Over 13 million ha of community pastures were devoid of vegetation due to over-grazing by 450 million heads of livestock. Due to the non-availability of wood on village common lands, rural women were compelled to spend 15-35 hours every week in walking long distances for collecting fuelwood from interior forests. As a result of deforestation, the damage from floods affected 58 million ha of agricultural land and over 60 million people during the 1980s. The extent of damage had increased by four folds over the earlier two decades.

The rural poor, particularly the women, who are primarily responsible for fetching water, fodder and fuel were faced with severe drudgery. The community felt that while it was the responsibility of the Forest Department to manage the forests, it was their inherent right to collect fodder, fuel and other products from the forests, without any obligation. Planting of trees for fodder and fuel was never considered by the farmers as a necessity. Meanwhile, faced with shortage of biomass, many village communities resisted the extraction of wood from the forests for commercial purposes. This led to the involvement of rural communities in forestry development programmes in India.

FORESTS AND COMMUNITY DEVELOPMENT

Because of their large number and wide distribution, the forest and wood-using industries influence the economic and social life of a great many communities. This influence is most keenly felt in the smaller cities and towns, particularly in wellforested regions, where they often constitute the dominant interest. In such situations, the prosperity of the community rises and falls to a very considerable extent with the prosperity of the forest-based industries. These have a decidedly stabilizing influence wherever the forests are managed on a sustained-yield basis, with the average annual cut and average annual growth in balance. Ghost towns do not occur where sustained yields are obtained and where mill capacity is adjusted to the growing capacity of the forest.

One of the important aspects of any industry is its influence on other industries and on service activities. The forest and wood-using industries purchase large quantities of axes, saws, tractors, trucks, planers, lathes, digesters, paper machines and other items of equipment; and they sell lumber, veneer, plywood, boxes, pulp, paper and similar items to other industries for use in the manufacture and marketing of their products. Railroads in some regions depend for their very existence on the freight revenue derived first from bringing raw materials, equipment and general supplies to communities dominated by one or more wood-using industries and secondly from transporting the products of these industries from mill to market.

Industry in general also supports many services, such as those supplied by grocers, druggists, clothiers, doctors, lawyers, ministers and teachers. The wood-using industries do not differ in this respect from other industries, except that in many rural districts they provide the sole or the main reason for the existence of a community. Where this is the case, they create the demand for such services; and where other industries are also present, they intensify that demand.

One further item of community concern is the relation of the forest and its dependent industries to the small landowners and especially to the farmer. The existence of a nearby wood-using industry provides a market, which in turn should encourage thrifty management of the wood lot. The additional income thus available can have a stabilizing effect both on the farmer and on the community of which he is a part. Forestry is a key to permanence.

FORAGE AND WILDLIFE

Not all forest industries are based on the utilization of trees. In many regions forage produced in openings in the forest and sometimes under the trees themselves, provides feed for cattle, sheep and pigs supplementary to that which they obtain from pastures and the open range. Beef, mutton and pork thus become in part products of the forest, the use of which for grazing may furnish an important means of support for the livestock industry. Timber production and livestock production are, however, not always compatible and stock must be kept out of areas where grazing is injurious to the forest. Browsing and trampling are likely to be particularly harmful in hardwood forests, from which domestic livestock should ordinarily be excluded.

Many fur-bearing animals which spend at least a part of their lives in the forest provide the raw material for another important industry. Milady must have furs; and mink, otter, fisher, marten, weasel, beaver, fox, muskrat, rabbit and many others supply them for her. Their production can well be one of the objectives of multipurpose forest management. The taking of other forms of wildlife by sportsmen and fishermen helps to provide a market for manufacturers of arms, ammunition and fishing gear. Users of the forest for recreational purposes step up business for makers of automobiles, cameras, clothing, insect-repellents and gadgets of all sorts; for keepers of service stations, hotels, motels and dude ranches; and for packers, guides and wranglers. This stimulation of economic activity may extend as far back as the iron mine, the petroleum well and the cotton field.

Recreation, Inspiration

The most important influence of the forest as a source of recreation is not indirectly through industry, but on the recreationist himself. Its chief significance lies in the enhancement of human values rather than monetary values. The forest provides not only recreation, but "re-creation."

In the hurly-burly of modern life, with its economic pressures and nervous tensions, people need occasionally to get back to nature. Where better than in the open and particularly in the forest, can they regain both physical and spiritual strength?

As civilization has advanced, man's relation to the forest has continually changed. To primitive man the primeval forest was generally a gloomy and forbidding place. He viewed it with mixed feelings of fear and superstition. Tree worship was common in many tribes. "The groves were God's first temples"; they were also the first temples of pagan gods.

Gradually, as tools were developed and new uses for wood were found and as the demand for land for agriculture increased, man pushed back the forest frontier. Domination of the forest by man was a slow and difficult process which developed qualities of courage, hardihood, resourcefulness and ruthlessness. The men who lived and worked in the forest and whose hero was Paul Bunyan, were no weaklings.

Today throughout much of the world the early relationship is reversed. Virgin forests, except in parts of South America, Africa and Asia, are a rarity. There is a shortage rather than a surplus of wood for industrial use. A rapidly increasing population is becoming more and more concentrated in urban centers. Even the lumberjack usually lives in town or in a comfortable camp quite unlike the crude shacks of earlier days. Forest areas of outstanding value for recreation are shrinking even as the need for recreation becomes more acute.

Different people seek re-creation of body, mind and soul in different ways. Some like to hunt and fish with gun and rod; others prefer to use field glasses and camera. It is now the wild beast that fears man, not man the wild beast. Some like to lose themselves on camping trips in an inaccessible wilderness; others are satisfied to enjoy the beauty of the forest from the seat of an automobile. Many wish enough "roughing it" to give them the illusion that they are emulating the prowess of their forebears. Some seek health, others inspiration. All are attracted to the forest by the opportunity to find rest or strife, relaxation or stimulation, in an environment that takes them out of the stifling routine of their daily lives.

From time immemorial the forest has influenced the work of architects, painters, poets, dramatists and preachers. It has molded the character not only of those who live and labour in it but of those who visit it only occasionally to play or to pray. In the realm of the human spirit its gifts are invaluable both figuratively and literally.As Robert Marshall expressed it: "The most important values of forest recreation are not susceptible of measurement in money terms.

They are concerned with such intangible considerations as inspiration, esthetic enjoyment and a gain in understanding. It is no more valid to rate them in terms of dollars and cents than it would be to rate the worth of a telephone pole in terms of the inspiration it gives." In face of mounting pressure for more wood for homes, furniture, magazines, rayon clothing and other tangible goods, it is important in the management of the forest to remember that "man doth not live by bread only."Forests are an inescapable geographic fact. Not only do they dominate the landscape throughout much of the world, but they influence man's life in countless and often unsuspected ways.

They temper the climate; improve the soil; reduce erosion and floods; support a wide variety of important industries; provide opportunities for the profitable employment of land, labour and capital; stabilize communities; and offer a unique source of recreation and inspiration in close communion with nature. Their management in ways that will assure the perpetuation of these values is an urgent task, of supreme importance to the people of the globe.

DEVELOPMENT OF COMMUNITY WASTELANDS AND DENUDED FORESTS

Conservation of biodiverstity and protection of forests are the other important components of community forestry. However there has been a lukewarm response to these activities in the past, mainly because of lack of awareness, inadequate benefits and lack of trust between the local communities and government agencies who own these lands. Hence the new strategy should find a suitable solution for these problems.

Development of forest lands: Rehabilitation of denuded forest lands and protection of natural forests are essential to conserve biodiversity. The forests influence the climate and soil moisture conditions and improve agricultural production.

This aspect needs wider publicity in rural areas and the local people should be motivated to take active part in Joint Forest Management. The Village Forest Committees willing to protect the natural forests, should be encouraged to share the benefits in the form of fodder, fuel, timber and NWFP. These committees can also take up reforestation of degraded forest lands, aiming to conserve biodiversity and generate income. While facilitating natural regeneration, the local communities can also introduce fuel, timber and NWFP species of their choice to enhance the productivity on denuded forest lands.

Development of community lands: Lack of people's initiative was a serious problem in developing community wastelands, mainly due to lack of direct benefits. Earlier efforts to establish woodlots could not generate substantial income. Establishment of NWFP species such as tamarind (*Tamarindus indica*), Indian gooseberry (*Emblica officinalis*), neem (*Azadirachta indica*), Pongamia (*Derris indica*), *Madhuca spp*. and a wide range of medicinal species can generate substantial income, regularly for over 50-100 years. With the world's herbal medicine business likely to cross US $ 3 trillion during the next decade, it will be an excellent opportunity for the Asian farmers to enhance their income through cultivation of these species on community lands as well as on private wastelands.

Development of Mined Areas: Rehabilitation of mined areas is another cause of serious concern. Although the industries are obliged to establish a green cover after mining the area, the outcome has been very poor. Such sites are major causes of soil erosion and flooding of rivers. Lack of commitment, inadequate expertise in selection of plant species and their management have been the main reasons for failure. These activities can be entrusted to voluntary organisations and local communities, with research support to raise healthy plantations. Preference should be given to NWFP and timber species which have long gestation period, sufficient to enrich the mined lands.

CO_2 Sinks: There are opportunities for environmental protection by establishing industrial greenbelts, recycling treated effluent and developing carbon dioxide sinks, particularly around urban areas. The polluting industries can be compelled to establish greenery through the involvement of voluntary organisations and local people.

Environmental Awareness and Eco-tourism: Lack of awareness among the common public is a cause for environmental pollution in the developing countries. Hence, continuous efforts are needed to motivate the common public to plant more trees and protect the environment. School children and literate urban population are fairly receptive to participate in environmental protection. This can be undertaken through school based plant nurseries, development of arboretum, botanical gardens and greenbelts around the schools and industries.

While developing industrial greenbelts and afforestation on river banks, special efforts can be made to promote eco-tourism. Creating waterbodies through plugging of gullies, developing, nature trails, shady grounds for resting, selecting suitable plant species to attract wildlife and providing adequate eco-tourism. Environmental exhibitions at these sites can help in motivating the public to take active part in community forestry and environmental protection.

ORGANISATIONAL DEVELOPMENT

Development of local organisations is the key to ensure people's participation and to sustain the benefits of investments and efforts made in community forestry.

Tree Growers Organisations

The initial need is to provide leadership to organise local people in afforestation. This can be carried out through formation of local Selp-Help Groups, village level farmers' associations and forest groups. This activity should receive priority during the next decade.

Capacity Building

Training of local farmers, supply of critical inputs, credit and technical advisory services to improve the profitability and developing market networks are the critical inputs which have to be introduced in the field to support tree growers.

Networking of Tree Growers and User Groups

There is an urgent need to set up networks of tree growers, forestry users and concerned scientists in these fields. Such networks should identify the opportunities for tree growers. They can also promote processing and marketing of forestry products and protect the interests of tree growers.

Forestry Information Service

Setting up of information centres to identify the priority of the industries and farmers and provide necessary information to enhance the economic viability of community forestry would help to sustain the interest of tree growers. Forestry Extension Units should closely associate with Information Service Centres to identify appropriate technologies for transferring to the field.

Forestry Research

The networks should identify the problems affecting the production and marketing of the produce and encourage the Forestry Research Institutions to take up relevant scientific studies to solve the problems. The researchers should closely associate with the network and Information Centres to disseminate new technologies to solve field problems and enhance the production.

Forestry research should focus on the following topics:

- Conservation of biodiversity;
- Improving the productivity of wastelands, tree plantations and natural forests;
- Exploring the utility of multipurpose tree species and NWFP;
- Developing of alternatives for conserving wood;

- Selecting various organisational systems to manage community forestry.

The outcome of these research topics can further strengthen the community forestry programme in the country.

EFFECTS OF SCALE ON COMMUNITIES OF INVERTEBRATES IN FORESTS

Many of the processes that govern the dynamics of insect associations with trees. In the main, these factors combine to keep insect populations at levels that do not appear to pose a threat to the health of trees or their commercial value. As Crawley (1989) has pointed out, there is a common perception that because the terrestrial environment is mainly green, then neither the plants nor their associated herbivores are limited in their resource requirements. However, the previous discussion has shown clearly that there are complex interactions between the different trophic levels associated with trees so that the end-result may give rise to major fluctuations in the relative performances of the different components. In both their natural and managed states, forests are affected by factors of scale such that regional site characteristics and local climate will influence the species mix and the growth rates of trees. Such effects will also influence invertebrate associates, which will respond both to resource availability (at both the plant and higher trophic levels) and to temperature and other abiotic conditions that will drive the rate of response. Some of the influences of scale are implicit in the species-area relationships that, at least in part, provide descriptors of species accumulation of both animals and plants in a particular region.

What, then, are the effects of scale on communities of invertebrates in forests and how are they related to forest management in a rapidly changing world? It has already been stated that the area occupied by a particular tree species is an important determinant of insect diversity, particularly if that area contains species that have been present for long periods. Such a situation applies to the natural oak populations in the northwestern USA, where two of the most convincing demonstrations of the species-area effect and interactions with other local and landscape features apply. The natural distribution of oak genera and species in this region covers a wide latitudinal and altitudinal range, including the Californian region (southern Oregon to northern Baja California bounded by the Rocky Mountain and Sierran regions, thus isolating the zone) and Atlantic region (larger, ranging from southern Canada to the Gulf of Mexico, west to the Great Plains). The oak species in the two regions are distinct from each other, those of the Californian region being closely related to Asian rather than eastern US oaks. Two separate studies of herbivores within specific feeding guilds have been carried out in these regions. Cornell and Washburn (1979) studied the Cynip-inae (Hymenoptera: Cynipidae), a subfamily of gall-forming wasps that are extremely specialized to the genus *Quercus.* They have been extensively studied and the data were used by Cornell and Washburn to investigate relationships to the areas occupied by oaks in the Atlantic and Californian regions. An earlier study by Opler (1974) also assessed relationships to oak distribution but was concerned with leaf-mining Lepi-doptera (Lepidoptera: Gracillariidae) within California only. Both studies demonstrated significant species-area relationships against total range (expressed in square miles). However, the pattern of response was quite variable depending on the subsets of data examined; some of these are reproduced in Table. The steeper slope and higher coefficient of variation in the Californian subset was attributed to the greater ecosystem variability in the latter location, taking account of the larger number of oak species, the wider altitudinal range and the greater variety of topographical features compared with the Atlantic region.

Interestingly, the leaf-miner data for the same geographical range of oaks suggest that new moth species colonize sympatric oak species within the same area and that there is no strong link with taxonomic relatedness between oak species, thus demonstrating a response more to total resource availability than to oak species *per se.* Such data provide support for habitat heterogeneity as one of the principal factors driving species-area relationships. Further studies by Cornell (1986) indicated that both the size of oak trees and, particularly, the size of the pool of cynipine wasp species in a given area were the major determinants of species richness on a given oak species.

He attributed this to several of the studies, although the wide variation in responses indicates that the effects may be tree and site specific. Elevated temperatures will also act on performance of both plant and invertebrate, changing both absolute growth rates and, perhaps more importantly, the synchrony between host plant suitability for feeding and presence of the appropriate stage of the insect. Those insects with overwintering stages that require quite precise coincidence of emergence with budburst of their host trees may be particularly vulnerable to changes in the incidence of frosts, especially if emergence is driven more by daylength than by temperature *per se.* Diapause termination, triggered by a specific period of chilling below a subzero threshold temperature, may also be affected so that emergence in the spring may be reduced, thus affecting overall population size. However, some insects presently constrained by low winter temperatures may show enhanced survival and increased reproduction with an increase in warmer winters. This is likely to be the case for green spruce aphid, *Ela-tobium abietinum,* which is normally killed by winter temperatures below -8°C.

Pollutants, such as ozone, SO_2 and NO_x, may have direct effects on tree performance and both indirect and direct effects on insect growth. Ozone results in amino acid changes within trees that can provide a richer food source for phloem-feeding insects in particular, potentially leading to higher growth and reproductive rates. However, recent studies by Holopainen *et al.* (1997) indicated that ozone episodes were less important than increased levels of nitrogen fertilizers in determining the performance of the aphids *Schizolachnus pineti* and *Cinara pinea* on Scots pine and *Cinara pilicornis* on Norway spruce. These authors concluded that occasional episodes of ozone may increase amino acid availability but, generally, the effects from predicted increases in atmospheric ozone were unlikely to be significant.

In considering effects on herbivores as a whole, several studies have demonstrated the high sensitivity of sap-sucking insects to the effects of atmospheric pollutants such that elevated levels of SO_2 and NO_x give rise to greater RGR and hence reproductive success, for example *E. abietinum* responded to increased concentrations of SO_2 with increased RGR when exposed on Sitka spruce. Foliage-chewing insects have also been shown to respond to atmospheric pollutants but with rather inconsistent trends. Scots pine seedling trees subjected to elevated SO_2 had higher sulphur, decreased calcium and higher levels of free amino acids and were shown to reduce the larval performance of pine looper moth, *Bupalus piniaria* (Lepidoptera: Geometridae), but had no effect on pine lappet moth, *Dendrolimus pini* (Lepidoptera: Lasiocampidae). The authors attributed the different responses to major reductions in glucose and fructose, which adversely affected *B. piniaria* whereas *D. pini* was able to tolerate the lower carbohydrate levels. The foliage of mountain birch, *Betula pubescens* ssp. *czercpanova,* subjected to experimentally enhanced levels of sulphur and nitrogen over a 10-year period had no effect on autumnal moth, *Epirrita autumnata* (Lepidoptera: Geometridae), feeding on the foliage.

Overall, despite high variability, it would appear that atmospheric pollutants have a neutral to positive effect on insect performance, although there are interactions with temperature, nutrient levels, etc. that can induce different effects even within the same insect species. Even when insect performance is impaired (*e.g.* elevated CO_2), the net effect may be increased damage as feeding rate increases to compensate for reduced food quality. This complex area of herbivore-plant dynamics clearly requires further work to determine patterns and causes.

FACTORS INFLUENCING PROCESSES AND OPPORTUNITIES FOR MANIPULATION

The array of processes considered in the early part of this chapter range from those reflecting different niches within a single tree, through local scales to the ultimate position of trees as major landscape components. It is also apparent that there are complex multitrophic interactions between trees and their herbivores that make it difficult to generalize concerning potential to manipulate processes. Management at the tree or stand level will involve local processes that reduce some of the complexity, such that gross climate effects may be removed as driving variables, although even here there may be microclimate effects at very small scales. Effects at this scale are manifested through the intrinsic suitability of the local population of trees, thus bringing in structural, nutritional and defensive qualities as well as the local impacts of natural enemies.

The interactions with wider-scale attributes. Extrinsic effects act at a range of scales, from local (within the forest), through regional to geographical, such that the effects on a given tree population become less but the wider attributes of climate, topography and site characteristics come into play. This will have the effect of changing the absolute and relative abundances of tree species and hence their associated invertebrate species over wide areas.

The examples of leaf-miners and gall-wasps from the full latitudinal range of oak species in California serve to illustrate this principle. In summary, therefore, any population of herbivores associated with trees is actually influenced by the herbivore-tree interface so that, provided the herbivore is able to find the host tree, its ultimate performance is driven by local factors that may in turn be influenced by the interaction parameters. Management at this local scale will reflect the initial components of tree species choice and location (*i.e.* the suitability of the site for tree growth and ultimate aim of the crop).

This, of course, applies particularly to artificial forests rather than natural forests, where tree species mix and stage are entirely driven by site and local climate characteristics. Once trees are established, the elements of further management that can influence susceptibility to herbivore colonization are addition of fertilizers, particularly nitrogen, and thinning regimes that will affect proximity to other trees and plants and generate open spaces and habitat corridors. All other factors will be extrinsic (proximity to other forests, topography, climate, etc.), with decreasing influence with both distance and time.

The influence of given factors is therefore a question of scale. If these are at the landscape scale, then it is possible to view forest management at a more remote level so that a high degree of fragmentation at a number of area scales can be considered and thus their relative influences on local herbivore-tree interactions measured at any fixed point within the wider landscape. This conceptual approach is now being used to increase understanding of the wider processes affecting invertebrate interactions with trees, making use of geographic information systems (GIS), new geostatistical methods and improved invertebrate sampling regimes to assess populations over relatively wide areas.

Further impetus to such approaches has been provided by the various biodiversity conventions and the implementation of biodiversity action plans at individual country levels. The principles described in this chapter apply equally to conservation of invertebrate diversity and abundance, thus embracing enhancement of biodiversity and avoidance of invertebrate outbreaks within the same conceptual framework.

Forest Fragmentation at Local and Landscape Scales

Among the parameters, the degree of fragmentation of a forest is one of the key variables that can be subject to human influence, both deliberate and accidental. It is also one of the variables most subject to natural disturbances over recent and geological time-scales. There is increasing interest in fragmentation in relation to tropical forests because of the extremely rapid loss of habitat through intensive logging. For example, models have been constructed to take account of the degree of fragmentation of tropical forests in relation to core area, edge effects and the shape of the forest area. Further refinement, including provision of spatiotem-poral components and the degree of connectivity between fragments, has been provided by Didham (1997). This synopsis included:

- Area effects, taking account of total area and interior core area which is not influenced by any edge effects;
- Edge effects, especially the distance over which the edge has an influence relative to the core;
- Shape of the fragment, which clearly affects both edge and core elements;
- Degree of spatiotemporal isolation so that the time since fragmentation is taken into account;
- Degree of habitat connectivity in the landscape, thus providing corridors for movement of species through otherwise unsuitable habitats.

Studies of the relative influences of edge to core in forests have been carried out in both temperate and tropical forest ecosystems, using a range of sampling methods. In general, it has been noted that invertebrate densities, of both individuals and species, tend to decline with distance from a forest edge to the core. Thus various beetle families show greater species richness or abundance at the edges of forests: Carabidae, Cerambycidae, Staphylinidae and mixed families of beetles.

Although not sampled in relation to direct interactions with host trees, the bark beetle *Ips typographus* tends to colonize edges more frequently than forest interiors. Patterns of abundance from edges to core woodland have been assumed to be a smooth gradation but recent studies on arthropod communities in Amazonian forests by Didham (1997) have demonstrated a bimodal form to relationships of edge to core abundances of leaf-litter invertebrates. This effect was compounded with the area of the forest fragment being studied, leading to three-dimensional relationships between fragment area, distance from forest edge and abundance of selected beetle species. Within the Amazonian forest system, the effect of fragmentation was felt even within large remnant forest blocks so that an area considerably greater than 100ha, possibly approaching 1000ha, was required before the invertebrate fauna matched that of undisturbed forests.

The implications of results such as these are difficult to extrapolate to the heavily managed forest landscapes of Europe and, increasingly, North America and elsewhere. Forest fragmentation can now be regarded as the norm in such situations and therefore the invertebrate faunas in such fragments may already be greatly impoverished compared with those in original undisturbed forests. However, the general principles of maintaining the ratio of edge to core to ensure that the habitat requirements of forest specialists as well as edge colonists are maintained is a sound strategy. Unfortunately, the biological attributes of edges and cores are not clearly defined and thus the type of edge may be as important as the edge itself. For example, studies by Ozanne *et al.* (1997) in canopies of managed conifer forests in Britain showed a lower species richness of various invertebrate taxa at the edges compared with the cores of forests studied.

This apparent contradiction to the studies cited earlier may reflect the type of edge, which in managed conifer plantations is often a very abrupt transition from the surrounding open ground area. The authors suggested that a gradation of the edge, with scrubby transition zones, could give rise to greater species richness at the edges but this would probably be at the expense of forest canopy specialists. Such uncertainty reinforces the need for more detailed studies of the mechanistic basis for observed effects of forest fragmentation, particularly taking account of the wide range of ecological factors that have been discussed in this chapter. Studies such as those of Waltz and Whitham (1997) provide useful data on the characteristics of individual trees that may influence edge and core populations of particular herbivores. They studied the effects of canopy structure in clonal poplars in relation to arthropod communities dominated by the gall-forming aphid *Pemphigus betae* and the leaf-feeding beetle *Chrysomela confluens*. There were direct correlations between the mature ramets of the trees and invertebrate species richness and this varied with the clone studied. The authors concluded that such within-tree variability in resource availability explained the higher biodiversity in mixed-aged stands and at edges compared with interiors of even-aged stands.

As Stork *et al.* (1997b) have pointed out, the understanding of forest arthropod dynamics, especially canopy-dwelling arthropods, is hampered by the tendency to gather observational data rather than by rigorous hypothesis testing. They also correctly point out the sheer difficulty of carrying out such studies, graphically illustrated by the 2-year effort by over 20 taxonomists at the Natural History Museum, London to sort and partially identify samples collected over a mere 12-day sampling period in Borneo. The underlying ecological principles that govern invertebrate communities of forests are clearly complex and yet unless they are incorporated into studies of forest fragmentation, there is little prospect of developing a sound science of both pest management and invertebrate conservation for the future.

Forest Disturbance as a Factor in Invertebrate Dynamics

The processes described above emphasize that size and scale matter in relation to communities of forest invertebrates. It was also stated that the areas of natural forest remnants are progressively shrinking and this is likely to have an impact on associated invertebrates. Much of the disturbance that leads to this extensive fragmentation is a result of human activity, directly or indirectly. Anthropogenic influences such as aerial pollution and climate change are having an increasingly global effect that will take many years to influence trees and their invertebrate communities. Prediction of these effects is imprecise because of the relatively poor understanding of the processes involved and the difficulties of tracking changes against an imperfect baseline.

Some indication of the effects of habitat fragmentation have been provided in studies of the outbreaks of some of the more damaging forest pests and these serve to illustrate the complexity of managing change. Edge effects are well known in relation to initial infestations of several damaging bark beetles, particularly in relation to felling programmes that can provide breeding resources as well as opening the remaining forest to the risks of windthrow, etc.

Ips typographies, the most damaging bark beetle in western Europe, is sustained at low densities in most forest locations by reliance on sporadic availability of weakened trees for initial breeding. However, if a sudden increase in breeding material is brought about by felling programmes or wind damage, populations can build up rapidly and attack surrounding living trees. Such eruptive patterns of attack are common in different species of bark beetle and the triggering of the infestation is frequently linked to large-scale provision of suitable breeding material on-site, either through human intervention or via natural effects such as fire or severe winds.

A good example of such an effect was the major fire in Yellowstone Park, USA during 1988 which left a mosaic of fire-killed, fire-damaged and undamaged areas in the area. Surveys showed that 12.7% of Douglas fir *(Pseudotsuga menziesii),* 17.9% of lodgepole pine *(Pinus contorta),* 6.6% of Engelmann spruce *(Picea engelmannii),* 7.5% of subalpine fir and 2.8% of whitebark pine *(Pinus albicaulis)* were killed following the build-up of bark- and wood-boring beetles in fire-damaged and stressed trees. The unexpected and severe nature of the Yellowstone fires also illustrates one of the adverse consequences of fire-management regimes that have been in place in the USA since the early 1900s.

Fire was previously a normal component of forest ecosystems in the USA and Canada and had a considerable influence on species composition, succession and forest stability. Regimes of fire suppression have changed the structure of many forests and, in some cases, have led to insect outbreaks.

A particularly good example of such an effect is the balsam fir and white spruce forest ecosystem of eastern Canada and the north-eastern USA. Fire suppression was started in 1920 and has been accompanied by increased frequency, intensity and scale of spruce budworm, *Choristoneura fumiferana* (Lepidoptera: Tortricidae), outbreaks.

Apart from any direct effects on the budworm, fire suppression altered succession by reducing pioneer species such as aspen, birch, jack pine and black spruce, thus decreasing the species mix that previously broke up the highly susceptible fir and spruce-fir stands. Recent strategies have been to return to selective use of prescribed burns to help restore the previous forest mosaics and thus to help combat spruce budworm outbreaks. McCullough *et al.* (1998) cite several other examples of linkage of insect outbreaks to fire suppression regimes and conclude that an improved understanding of the ecology of fire-tree-insect interactions will help to identify strategies to manage pests and to maintain biodiversity in northern forest ecosystems.

CHALLENGES OF PERFORMANCE MONITORING

Performance monitoring is a key theme as it relates to an ongoing need to accommodate a complex array of monitoring, auditing, validation and quality control requirements over a range of multiple objectives. The challenge in this realm is that capacity of local community-based systems is often stretched to the maximum in performance monitoring, and interests of different stakeholders may not be fully aligned.

As an example, many carbon projects now have multiple objectives – carbon, biodiversity, and community welfare – but there are few received protocols for addressing these concurrently. The main strategies for addressing community based monitoring challenges rely on: (i) training; (ii) local innovation; and (iii) development of simplifying procedures. In Senegal, plantations are operated by direct planting of seedlings without any nurseries; monitoring activities are conducted at the community level and focus on aspects such as plantation of the large-scale project and taking care of the young planted seeds.

A key monitoring challenge has been to make communities understand the key factors of good plantation performance and then to target incentives to these very factors. It was resolved finally through the same factors which are described between Livelihoods and the project developer: "GREEN" coded areas (corresponding to good growing conditions) with the right tree density indicated good performance. A second challenge specific to monitoring included stratification and how to monitor the evolution of the tree density and then measure the growth of the trees. This was resolved through combining a mix of external expertise informed by local observations, similarly to how soil quality was monitored. A key lesson from this was to choose from the outset the right "community-based" performance incentives, and be certain that these factors are included in the common contract you have with the project developer.

Monitoring challenges are significant in the Indian Sundarbans. Destructive sampling is illegal under national laws because mangroves are protected species. Three months of the year the area is prone to storms that routinely take lives and damage property.

Loss of life to tigers and snakebite is not uncommon in the region: during the first planting season a villager gathering propagules for the project was attacked and killed by a tiger. Satellite imagery and remote sensing are confounded by high tide levels which submerge plantations; only multiple views might give some idea of plantation extent or condition and these are too costly to purchase given the budgets of most of these projects. Traditional forest mensuration methods work at early stages of plantation growth but once a plantation is 5-10 years old it is impossible to move through and measure safely. Errors are potentially large, and the diversity of species escalates from natural regeneration once the initial plantation has taken hold. At present, all of the plantations are readily measured but the situation is expected to become more difficult. Work has commenced with the Sundarbans Forestry Department to establish protocols that can more readily assess above ground biomass; estimating below ground biomass and soil carbon remains elusive and the project will rely on default methodologies and calculations for working estimates. In the future, destructive sampling may also be permitted to obtain better estimates, but carbon measurement remains difficult under these conditions.

Although carbon monitoring can follow certain standard protocols, monitoring community benefits and biodiversity in the Indian Sundarbans is not straightforward. Criteria call for monitoring of indirect and direct impacts, but there are no standards for what might be regarded as positive or negative effects, beyond simply following locally developed preferences. These seem to us somewhat arbitrary, and in any event are more useful as a feedback device for ongoing management, rather than a definitive metric for defining success. Much of the problem deals with measurement of the baseline and specification of counterfactual conditions; livelihood activities can be wiped out in one fell swoop by a bad storm, or by a pest, through little fault of the communities.

A diversity of plantation activities might make them more resilient to future losses, but there is no way of knowing how well they would have done in the absence of the activities. Moreover, most monitoring tends to be input oriented (*e.g.*, people involved, hours spent) rather than results oriented (*e.g.*, income levels achieved, health and nutritional status improved). Given that results are often long-term, monitoring timeframes and impacts are more difficult to tie to specific project activities. Biodiversity impacts fall in a similar realm: the timescale of the impacts is not in concordance with that of most short term monitoring protocols. At this stage, there has been no particular need for detailed impact monitoring in the Indian Sundarbans, but this may change in the future.

In the India project with Naandi, a key monitoring challenge has been to collect vast amounts of data (area, species mix) for a single plot, to treat these, and to store the information knowing that carbon project eligibility procedures require saving and being able to trace single items from year 0 to year 20. It was resolved primarily through a tool developed by Naandi called "Livelihoods 360" relying on the mobile phone network. From a standard mobile phone farmers or field officers, can collect data which are sent to central servers were the data are treated; the majority of farmers – even the poorest – in these communities now have a mobile phone. This also allows the project developer to send back treated information to the field officers and farmers. A second challenge specific to monitoring included limited skills of farmers, who more accustomed to managing cattle than to managing fruit trees. This was resolved through capacity building in centers such as the central nursery, and it will be developed further.

Many of the general issues faced by others – as noted above – are also faced by Yagasu in Indonesia. But some of Yagasu's approaches show how innovation can potentially lead to better performance monitoring. The issue of below ground biomass measurement is problematic worldwide; few empirical studies have been done and results are frequently not easily transferred to other locations because of differences in biophysical conditions. From the outset, Yagasu has maintained a very high level of research interest and capacity on issues associated with carbon measurement and tree growth. At the proposal stage, for example, staff had been working on fractal models of tree and root growth, which would assist in estimation of below ground biomass using non-destructive methods. Since then, ongoing research programmes on BGB will assist in estimates of carbon sequestered by the project. The work can assist projects in the country by providing a benchmark national value for use in those estimates.

8

Model Social Forestry Programmes

Few models of social forestry progarammes are presented here. These case studies provide some insights into the way they have been planned and implemented. Bharatiya Agro-Industries Foundation, Urulikanchan, Maharastra: The Mahatma Gandhi Vidyalay, started in 1950, has been active in promoting afforestation. More than 2,750 students planted one lakh subabul trees for fodder and other uses in 1981. Eighty per of these have survived. Encouraged by this, they planted about 2 lakh saplings more in and around their farms and farmstead in 1982. The experiment shows the potential the school systems hold in strengthening social forestry programmes.

Experimental social forestry in Garadgaon, Maharastra: Dr. Arvind G. Raddi and his colleagues prepared a social forestry scheme to make Garadgaon self sufficient in its fuel and fodder requirements. For this they utilized the funds available under the employment guarantee scheme of the state government. Through an informal survey they identified the following.

- Total boundaries of the village land under cultivation was more 1,50,000 metres where 23,150 trees could be planted. Fodder grass could be grown in the space between these trees.
- The village stream is 2.67 km and about 26,700 trees could be planted on its banks.
- As many as 3,330 trees could be planted along the road to 1.67 km. A total of 3,730 trees could be planted if other small roads were also taken into account.
- A nursery could be raised near the perennial stream.

The total budget of the scheme was about Rs.4.25 lakh. Dr. Raddi estimated that 241.5 tones of fuel wood would be required every year (805 tree) for the whole village. Similarly, fodder requirement was also calculate. He thought of establishing a fuel wood depot and a fodder bank in the village to be managed by the gram panchayat. The progamme was implemented in 1979. By and large, the results were quite encouraging. This experiment successfully combined private initiative, proper use of the government wasteland and administrative resources for the community benefits.

Government of Orissa

Orissa has a forest area of 38.44% of the total geographic area of the state. Although this is higher than the average forest area of the country, the rate of deforestation is also higher in the state. In 1983, Orissa Government in collaboration with Sweden started implementing the social forestry progamme in the state.

The aims of the programme were:

- To meet the fuel, fodder and other domestic needs of the rural population,
- To restore the ecological balance,
- To check soil erosion, conserve soil moisture and increase the soil fertility and
- To meet the needs of forest based industries.

The project had the following targets:

- Establishment of village plantation,
- Rehabilitation and replantation of degraded forests,
- Stimulation of private forest plantation through various assistance, and
- Exploring new possibilities for "Forest Farming for Rural Poor".

The social forestry programme, started in four districts in 1984-85, was extended to another five districts in 1985-86.

The evaluation of the project shows that the achievements were much higher than the targets. The project was implemented in 5000 villages helping the people meet some of their basic needs improve the quality of their life. It inspired them for collective community management and sharing the benefits among the poor.

Gram Vikas, Kolar, Karnataka

Gram Vikas is a voluntary organization set up by Dr. Arentha Iyer in the Kolar district of Karnataka. He motivated the children to plant trees adjacent to the school or balwadi. Following its success, men and women were urged to join the afforestation programme and nursery raising in a big way.

They decided to raise a nursery for:

- Farm forestry of private persons e.g the Neyveli Lignite Corporation,
- The forest department for their distribution purposes,
- Local requirements for planting trees under community woodlots programme,
- Farm forestry-Casuarina and Eucalyptus
- Community plantation- coconut, mango, jackfruit, medicinal plants, subabul and other trees for fuel wood; and
- Fruit yielding trees, flower plants and trees for fuel wood near their houses.

Gram Vikas extended help to the people in the form of:

- Initial capital
- Technical guidance
- Organizational support
- Animators and
- Introduction to various government officials
- The organization had to face several constraints such as:
  - Lack of support and cooperation from the forest department,
  - Lengthy government procedures involved in the land distribution to people for affoestation
  - Shortage of polythene bags, and
  - Scarcity and poor quality seeds.

In the final analysis the success or failure of a project depends on its economic viability. From this stand point, this programme has been successful. The cost of production under this programme was Rs. 11 per 100 seedlings, whereas the sale price was Rs. 15 per 100 seedling in 1984-85 and Rs 20 in 1985-86. For the success of social forestry, it is essential to have a joint and collaborative effort of all the people. In the absence of assured support from the political, bureaucratic, panchayat and the local people social forestry cannot achieve its objective.

However, direct economic benefit is the key to people's participation in social forestry programmes. There is no substitute for people's participation to make social forestry programme a success as proved by the pioneering projects in Gujarat.

SOCIAL FORESTRY PROGRAMMES FAIL

It is a fact that many social forestry programmes have failed miserably. Instead of presenting case studies, an attempt is made here to identify the common factors that lead to in the failure of social forestry programmes. Special effort is made to explain each factor separately for emphasis though some of them appear to be overlapping.

- *Lack of involvement of people:* Often social forestry is started by the initiative of some persons without conscientising and motivating the people on the need to cooperate in it. In such cases people do not feel that the plantation is theirs nor do they feel responsible to look after it.
- *Absence of requirement and resource survey:* Plantation is carried out without proper survey and assessment of the local resources (land, local species, labour, marketing, facilities etc.) and requirements of the people in terms of firewood, fodder, small timber, fruits and manure. Instead of selecting species according to the requirement of the people and agro-climatic conditions whatever available is planted. Hence social forestry becomes irrelevant to people.

- *Lack of awareness:* There is a strong feeling among people that planting and caring of trees does not require any technical know-how. People say, " Just dig a small hole in the soil, plant the seedlings or sow the seeds and they will come up by themselves. You see, in the forest no one cares for the trees and yet they come up." Theoretically this argument sounds convincing. But practically it is not true. Saplings planted in social forestry areas do require care and protection like small children. They have to be looked after and protected regularly at least for five years. Several operations should be carried out one by one during the first three to five years period after plantation.
- *Improper method of planting and caring:* Method of planting influences greatly the survival and growth rate of saplings in a social forestry programme. The size and depth of pits and distance between them, planting design 9square, rectangular, diagonal, hexagonal contour and random planting, size of seedlings, depth of planting soil and moisture conservation measures, weed, pests and disease control have to be planned and executed according to the soil, climate, rainfall, topography and location of the area. The only method of planting people adopt is to dig a hole and plant whatever saplings available.
- *Lack of preparation:* Proper planting requires sufficient preparation and planning. Pits have to be dug moths earlier if not an year ago, seedlings should be raised at least one year in advance according to the species and number required. Social forestry cannot succeed without planning and preparation.
- *Lack of target orientedness:* Most of the social forestry programmes are implemented as a social even once a year. Every year 'Vana mahotsava" (festival of forests) is celebrated mobilizing as many people as possible to pant as many trees as possible. It is celebrated with pomp and gaiety and is used as an occasion to politicize and socially organize people. The actual implementation part of social forestry is completely neglected. The result is that the social forestry area is left with empty pits. Every year "Vana mahotsava" is celebrated and saplings are planted. Even the prestigious Wasteland Development board, which was constituted to afforest 5 million hectares of wasteland per year, could not afforest one million hectare even in three years. Hence the remark: "Vana Mahotsava" is planting trees in the same pit every year.
- *Lack of technical education:* Raising a suitable social forestry is a technical operation which necessarily requires technical know-how. Social forestry is a planned plantation utilizing land, labour, capital and plant and animal resources. Hence the planning has to be scientific

and technically viable. To prepare a budget for a technically viable social forestry programme, one should have the technical know how without which he cannot make a proper financial proposal. Lack of proper technical know-how leads to:

- Poor planning
- Non-viable financial proposals (under or over financing),
- Improper utilization of resources such as labour, lands, water, plants and animals,
- Improper management practices and poor organization.
- Poor organization.

Surely such social forestry programmes will fail.

- *Planting too small saplings:* Normally one year old or one metre high saplings should be planted in the field. However, the practice is to plant very small saplings which may be only few months old. Those who raise the nursery do not like to keep the saplings up to one year because of additional expenditure and lesser profit.
- *Planting naked saplings:* Naked saplings are those with bare roots or without a mud ball in the polythene bad. These are saplings, which are pulled out from the nursery, and in that process several roots get damaged. Besides there delay of several hours between the removal of seedlings from the nursery and planting in the field. During such long intervals the naked saplings wilt without water and nutrients and these cannot survive in the field. It is important that the saplings used for social forestry should be grown in the polythene bags or be removed with a mud ball from the nursery. (Consult Booklet No. 5. "*Nursery Raising*" 5, FS-5 for details on Nursery raising of plants and Booklet No 6. "Propagation of Fruit Trees" OMS-4 for details of different methods of plant propagation).
- *Indiscriminate grazing:* To ensure the success of social forestry, cattle should not be allowed to graze till the trees have grown beyond their reach. Cattle can browse one-year-old saplings. That means the time, labour and capital spent during one year is wasted.
- *Exploitation:* When social forestry programme starts yielding, the rich and powerful of the place misappropriate the yield leaving the poor with nothing.
- *Poor or no marketing facilities:* People will not work for social forestry unless it is paying to them. Adequate remuneration for the work and the product is everyone's right. Unless marketing is organized and prices are stabilized it will be impossible to obtain adequate remuneration for every one in social forestry programmes.
- *Lack of a suitable administrative system:* Malpractices and exploitation by the rich and powerful in the distribution of social forestry products

occurs mainly because there is no alternative body authorized to handle the distribution. Each operation in social forestry such as protection, maintenance, harvesting distribution should be administratively streamlined and should not be left to the rich and powerful.

- *Lack of legal protection:* People who are involved in social forestry should get their rights. Sufficient legal safeguards should be ensured to protect the legal rights of the beneficiaries of social forestry programmes. Illegal occupation of land, misappropriation of social forestry products should be prevented by legislation. Clear laws should be formulated regarding social forestry.
- *Lack of vision:* Often social forestry is implemented as an end in itself for meeting the basic, immediate needs and to generate additional income. It should further move into village level processing social forestry products and agro-industries based on them. These in turn will give rise to other enterprises in marketing, storage and transport resulting in more employment opportunities for unemployed youth. Thus one should go beyond the mere physical implementation and management of social forestry.
- *Lack of soil and water conservation:* Like any other plants in the nature soil and water are the basic requirements for the growth of trees in a social forestry. Though conservation of soil and water one of the main objectives of social forestry programme, the successful establishment and growth of saplings up to five years require special attention to the soil and water conservation practices which are seldom included in the programme (Consult Booklets on Soil and Water Conservation Nos. 1 to 15),

FOREST POLICY

Before 1865, forest dwellers were completely free to exploit the forest wealth. Then, on 3 August 1865, the British rulers, on the basis of the report of the then-superintendent of forests in Burma, issued a memorandum providing guidelines restricting the rights of forest dwellers to conserve the forests. This was further modified in 1894. The sole object with which State forests are administered is the public benefit. In some cases the public to be benefited is the whole body of tax payers; in others the people on the track within which the forest is situated; but in almost all cases the constitution and preservation of a forest involve, in greater or lesser degree, the regulation of rights and restriction of privileges of users in the forest areas which may have previously been enjoyed by the inhabitants of its immediate neighbourhood.

This regulation and restrictions are justified only when the advantage to be gained by the public is great and the cardinal principle to be observed is that the rights and privileges of individuals must be limited otherwise than for their own benefit, only in such degree as is absolutely necessary to secure that advantage. In actual practice, however, all these pious declarations were set aside whenever they came in the way of British interests. For example, forests in Nagaland and the Terai were unscrupulously cut to meet the increasing demand of wood during both world wars.

The National Forest Policy of the Government of India (1952) is an extension of this policy. This policy prescribed that the claims of communities near forests should not override the national interests, that in no event can the forest dwellers use the forest wealth at the cost of wider national interests, and that relinquishment of forest land for agriculture should be permitted only in very exceptional and essential cases. The old policy of relinquishing even valuable forests for permanent cultivation was discontinued and steps to use forest land for agricultural purposes were to be taken only after very serious consideration. To ensure the balanced use of land, a detailed land capability survey was suggested. Conservation of wildlife was to be regularized. The tribals were to be weaned away from shifting cultivation. The concept of "national interest" has been applied in a narrow sense. A welfare state cannot have a basic contradiction between local and national interests. As the analysis that follows will show, in the implementation of the forest policy, "national interests" remained confined to augmenting revenue earnings from the forests. Whenever the interests of the local people or ecological considerations hampered possible revenue from forests, the forest department pushed them aside on the pretext of broader national interest.

Forest dwellers have been dissociated from the management and exploitation of forest wealth. The British contractual system that still exists in many states has resulted in unscrupulous exploitation of the local people and of the natural vegetation and wildlife that the forest policy was intended to conserve. Development programmes-construction of roads and availability of educational, medical and housing facilities-have allowed economically viable outsiders to enter forest regions. In order to make quick profits, they have exploited the forest dwellers, displacing them from their land and making them bonded labourers. Except for the states of Haryana, Punjab, Rajasthan, Tripura, Andaman and Nicobar Islands, Goa, Daman and Diu, the government has been earning huge net forest revenues. Over the years, this revenue has been increasing.

On the other hand, except in a few places, the condition of the forest dwellers has been deteriorating. The government's various development programmes and the tribal welfare schemes have by and large failed to make any dent in the deteriorating condition of the forests and forest dwellers. The crux of the problem lies in misdirected policy and its half-hearted implementation.

RESTRICTIONS AND BACKBONE ON FOREST DWELLERS

The backbone of forest dwellers' economy is the vegetation found in the nearby forests. But the forest dwellers' rights to collect fuel, fodder and minor forest produce are very restricted. (In Arunachal Pradesh, the tribals have special rights to collect all forest produce and hunt and fish freely in all forests, whether reserved or unclassified. This concession is not found anywhere else). In almost all states, forest dwellers can collect forest produce, either free or at concessional rates, only from protected and unclassified forests. Even these are not legal rights, but are subject to various government restrictions and regulations. These facilities can even be terminated if the capacity of forests does not permit the exercise of the right. The right holders are also obligated to help the forest department prevent and fight forest fires and to help the staff in connection with forest offenses. In West Bengal, since Independence, forests have been the bone of contention between the forest department and the forest dwellers, most of whom are tribals.

In forest dwellers contend that they, by virtue of being native people, have the right to use forest trees for their livelihood. One of their complaints is the cattle trenches dug around the forest areas prevent the free flow of water into their arable land. They were charged high rates for permits to collect minor forest produce, especially tussar cocoons. Grazing has been prohibited and their agricultural land has been recorded as forest land in the recorder of rights; later the tribals were asked to vacate the land.

The forest department, on the other hand, blames the tribals for wanton and indiscriminate destruction of wilds vegetation and wildlife, sand urges the restriction of their rights over the forests. The tribal welfare department, while conceding the need to preserve the forest, says that the tribals do have special claims over the forests and forest produce. A government study team found the attitude of the forest department in many states to be one of callousness, indifference and neglect toward tribals. For example, the forest department in Karnataka (then Mysore), in its enthusiasm to set up a game sanctuary in the state, drove out the local inhabitants without making alternate arrangements for their resettlement.

The intervention of the social welfare department was of no avail. The roads constructed in reserved forests by the social welfare department of the state for the benefit of the tribals, on the express understanding that they would be maintained by the forest department, have been badly neglected. The forest department must be persuaded with great difficulty to release water from tanks lying in reserved forests for irrigating the tribal areas lower down. In Andhra Pradesh, too, extensive areas under cultivation were being converted into reserved forests. The forest department has extended its boundaries almost to the doorstep of many tribal villages, instead of leaving open a strip, at least one kilometre in width, between the villages and the forest boundary. Areas were also included in the forests which had no tree growth at all and heavy fines were being imposed on tribal encroachers.

In Tamil Nadu (then Madras), the study team found discrimination against tribals but not non-tribals in Wenlock Down. The latter were given land on pattas (land titles) either permanently or for long periods, but the Todas-local inhabitants-received only annual cultivation permits. The study team also noted that extensive areas now classified as forests were under unauthorized cultivation. Without rights of ownership, the tribals could not undertake soil conservation measures. Moreover, the forest department's practice of serving eviction notices to tribal encroachers and fining them operated as a great hardship to the tribals. Finally, the study team observed that tree pattas, which entitle the holder to the use of the fruits of trees, were issued in Coimbatore district to non-tribals living far away from the tribal settlements, in respect to mango and tamarind trees in the tribal villages.

O.P. Arya, in his interview with the officials of the forest department, voluntary agencies and forest dwellers, has tried to show the "bitter relationship between the forest administration and the local populace." The divisional forest officer he interviewed was "critical about the attitude of tribals toward the forest." His view was confirmed by the collector and the charge officer. On the other hand, the people-about 80 percent-complained during the field survey about harassment by the forest officials. The privileged ones, however, had good rapports with the forest department.

The Irish missionary stationed in the area who, according to the author, gave "a very impartial view of the whole problem," said that "the forester used to take a lot of benefits from the tribals of the village. Some of them used to be employed in the forest and were paid much less than government rates. He used to take gifts in kind, too." Since the villagers depend solely on forests for their livelihood, the forester behaves like a ring-master and makes them do what he desires.

Forest Dwellers and Department

The various kinds of restrictions imposed on the forest dwellers virtually put them at the mercy of the forest department, especially lower-level functionaries. Illiteracy and poor economic conditions make them even more vulnerable. In some areas in Andhra Pradesh, the forest guards have their cut of minor forest produce earnings, and in some areas the tribals are often made to work without pay.

In spire of the rest houses spread all over the block, the forest rangers or high level functionaries could not find it convenient to inspect the area. This is not an isolated instance. The IAS probationers, during their survey of socioeconomic conditions of tribals in different states, noted different modes of exploitation by the forest guards and other lower-level functionaries who were invariably in league with the patwari and the local police constable. Forest dwellers have also become victims of the commercial exploitation of the forest. Since British rule, contractors have employed the forest dwellers to do unskilled jobs for low wages and in appalling conditions. In the late 1940s, Symington recommended the association of local inhabitants in the exploitation of forest produce. In 1946-1947, the Forest Labour Societies were initiated by B.G. Kher in the then Bombay state. The purpose of the Forest Labour Societies was "not only to give the Adivasi labourers full wages along with a share in the profit, but also to train them gradually to take up the responsibility of conducting forest and other business by their cooperative efforts."

The Bombay state experiment was complemented in the First Five Year Plan. The Second Five Year Plan recommended the establishment of Forest Labour Societies in an increasing number in all states. It recognized that "the manner in which the forest resources are exploited has a great deal of bearing on their welfare. In many ways the penetration of forest contractor into tribal economy has been harmful". The Third Five Year Plan stated that "development of forestry and forest industries is also essential for raising the income of the tribal people who live in the forest areas".

The Working Group of Welfare of Backward Classes for the Fourth Plan stated that the manner in which the existing forest policy is understood and implemented had placed the tribals completely at a disadvantage. The Resolution of the Central Board of Forestry reiterated replacing the contractors with Forest Labour Cooperatives. Gradually, the Forest Labour Cooperatives were set up in most of the states. But the contractual system of exploitation has not yet been rooted out. The Union Minister for Agriculture, Rao Birendra Singh, promised in the Parliament to end the contractual system of exploitation of forest wealth within three years. In fact, the forest dwellers' share as wage earners in cutting, logging and loading was significantly less than the ruling market price. The forest dwellers can gain from the forest wealth only when they are involved in its processing through cooperatives.

The forest department also employs labourers for a variety of tasks such as road repair, road making and silviculture operations. Workers are also employed for departmental construction-building bridges, culverts and causeways. This gives much needed additional income to the tribals. With the expansion and development of forest-based industries, the demand for the exploitation of minor forest produce has increased. Unlike the major forest produce, minor forest produce in several states is collected by the government agencies and in others by businesspeople. Both the forest department and businesspeople are exploiters. In Madhya Pradesh, trade in tendu, patta, timber, gum, harra, bamboo, sal seeds and khair is nationalized. The payment of wages in the collection of tendu leaves is on the basis of number of bundles collected. The rate per bundle is different in the bordering states of Bihar and Uttar Pradesh. For example, when in Uttar Pradesh the rate was higher by 20 paise

(US 20) per bundle, a person could not collect more than two bundles a day. Just to earn 40 paise (US 4c) more, the tendu leaf collectors in Madhya Pradesh used to walk 12-20 miles during the night and sell the bundles in Uttar Pradesh anon.

In Orissa, the forest department has taken over the collection of tendu leaves. The forest region is classified into subdivisions and ranges. Each range has about 10 collection centers from which pluckers receive 10 paise (US 1c) for leaves. In this way the government earned annually Rs 44 million (US $3.4 million) or royalty, and the Forest Corporation a profit of Rs. 20 million (US $1.5 million). But this profit had been made at the cost of the tribals.

Where the minor forest produce is not nationalized, businesspeople exploit the forest dwellers. In the Uttarakhand region in Uttar Pradesh, contractors employed local people at nominal wages, but made huge profits themselves. In Chamoli district the Dashauli Gram Swarajya Sangh intimidated the local people about the prevailing market prices of the different herbs and mushrooms they had collected, and advised them to sell their herbs rather than collect them at nominal wage for the contractor.

Some tribal regions have no regular market. At the weekly mandis (marketplaces), private wholesale dealers buy the minor forest produce at very low rates and sell them at huge profits in the cities. Many times they pay the tribals in advance and later buy their goods at nominal costs. Since most of the tribals do not know how to count, the price that is promised often does not correspond to the amount they actually receive. Generally, the trader is a moneylender, too, who buys the produce from the tribals as repayment of a debt, but at incredibly low prices. The Tribal Development Corporations, Forest Corporations and some other cooperative agencies have taken up the sale of some produce, but the moneylender-cum-trader has yet to be eliminated by them.

URBAN FORESTRY PROGRAMME

The 1990 Master Plan for Forestry Development (MPFD) included Urban Forestry Programme as one of the five component programmes under the umbrella programme "Programmes on Man and the Environment". It defined urban forestry (UF) simply as "a forest park, a nature centre, a boulevard lined with trees or flowering plants or even coconuts, a street with trees and other green plants in the islands or sidewalks, a vacant lot planted to trees and other green plants, a school or hospital or a factory yard with trees. In short, urban forestry is a greening movement, a people-oriented forestry designed to raise the quality of environment of the people in urban centers".

Land, water and air pollution and space congestion have become very critical in at least 4 big cities in the Philippines. These include Metro Manila, Metro Cebu, Davao and Cagayan de Oro. The rapid degradation and deterioration of the quality of the environment in these urban centers result to health hazards to the residents.

Everyday, large quantities of $CO_{2,}$ CO and other toxic gases are spewed into the air from factories, manufacturing firms, transportation, burning by residents themselves. This is aggravated by noise pollution from engines and horns of motor vehicles.

The influx of people from rural areas in these urban centers results to overcrowding of side walks, canals, railroad shoulders, underneath fly-overs, bridges and any conceivable vacant space. For instance, it is estimated that there are about 10M people in Metro Manila, 30 – 40% of whom live in squatter areas. The voluminous solid wastes generated by the residents, improper wastes disposal and the lack of basic sanitary facilities of squatter colonies contribute to the worsening health and sanitation problem.

In order to mitigate these environmental problems, establishment and maintenance of urban forests is one of the interventions sought. Among others, the trees and other green plants absorb CO_2 and gives off $O_{2,}$ hence they mitigate or lessen pollutants from the atmosphere. They also provide shade, serve as buffer against noise and air-borne dust and conserve water and soil. Ganapin (1993) comprehensively discussed the different roles and benefits of urban trees to environmental enhancement categorized under 3 different uses: climatological, engineering, and architectural/aesthetic uses.

GOAL, OBJECTIVES AND STRATEGIES OF UF PROGRAMME

The goal of the programme is the rehabilitation of the environment in urban centers. Specifically, as defined in the MPFD (1990), the medium term (10 – 15 years) goal is to attain active practice of UF in cities and towns.

Its objectives are as follows:

- To establish tree strips and forest parks to help in reducing air, noise and sight pollution and in improving air temperature in urban area and to attain a tree to person ratio of 1:4; and
- To contribute to physiological and psychological well – being of urban residents.

The following are the strategies to attain the goals and objectives of UF programme:

- Establishment of mini- forests and nature parks;
- Greening of main thoroughfares, side streets and islands;
- Establishment of city/municipal/school nurseries;
- Enlistment of participation of LGUs, NGOs and other government agencies in planting trees in school grounds, military camps, hospital compounds, etc;
- Provision of tax incentives to encourage owners of vacant and idle lots to plant trees;

- Initiation of an intensive information campaign on the benefits derived form urban forests; and
- Enactment of laws and local ordinance directed towards urban environmental enhancement.

Due to funding constraints, the MPFD initially targeted the following six heavily polluted and/or congested cities for UF development: Metro Manila, Cebu City, Davao City, Cagayan de Oro City, Iloilo City and Zamboanga City.

The components of the programme include the following:

- Urban forest and nature parks development - Mini-forests and parks are targeted to cover 1-5 ha per forest or park depending on availability of area. Depending on the population, the target is to establish 1 park per 100,000 – 150,000 residents at strategic locations in population centers. The species composition will be mostly forest trees with emphasis on ornamental, fruit – bearing (including feeding tree) and shade trees. Species will be selected based on soil, climate and resistance to pollutants.
- Grounds landscaping (compound planting) - This involves planting of trees and other plants in compounds/grounds of schools, government offices and hospitals, military camps, compounds of factories and commercial establishments including subdivisions.
- Greenbelt development (roadside planting) - This project involves greening of main thoroughfares, side streets and roads. It is to be a joint undertaking of LGUs, DENR and NGOs. In the preparation of overall urban forestry plan for each city, the main streets and roads to be planted will be identified.

The following support components should be considered in UF programme planning and development:

- Policy and legal reform - Zoning should be incorporated in the city plan (*e.g.* placing industries in city outskirts to reduce pollution within the city).
- Institutional strengthening - In order to oversee and coordinate the urban forestry programme in the city, an Urban Forestry Division should be created in the Forest Management Service sector at the regional or PENRO offices of DENR similar to the one created at FMS – NCR.
- Human resource development - Due to limited expertise and practical experience on UF in the country, there is a need to conduct training in UF especially for DENR and LGUs who will be implementing the programme.
- Research and Development - UF research is needed on: 1) effects of different pollutants on physiological processes, growth and survival; 2) species – site compatibility; and 3) care and maintenance techniques in an urban setting.

- Information dissemination - A public information campaign (*e.g.* through mass media; including teaching the values of trees as one of the topics in science subjects in elementary grades) is needed on the importance of trees in urban areas so people will be sensitized not to vandalize or injure the trees planted.

FOREST LEGACY PROGRAMME

The Cooperative Forestry Assistance Act (CFAA) of 1978, as amended, (16 U.S.C. 2101 et. seq.) recognizes that the majority of the Nation's productive forest lands are in private ownership; that private landowners are facing increased pressure to convert their forest lands to other uses; that greater population density, user demands and restrictions on Federal and other public lands are placing increased pressures on private lands to provide a wide variety of products and services from working forests including timber and other forest commodities, fish and wildlife habitat, watershed function and water supply, aesthetic qualities, historical and cultural resources, and recreational opportunities; and that good stewardship of privately held forest lands requires a long-term commitment that can be fostered through a partnership of Federal, State, local government and individual efforts.

In 1990, the Forest Legacy Programme (FLP) was established to promote the long-term integrity of forestlands. The Secretary was directed to establish the FLP in cooperation with State, regional, and other units of government.

In carrying out this mandate, the Secretary of Agriculture is authorized to acquire lands and interests in lands in perpetuity for inclusion in the FLP. Landowner participation in the FLP, including the sale of lands and interests in lands, is entirely voluntary. The Programme is implemented through State participation, consistent with these National FLP guidelines, and as described in each State Assessment of Need.

The FLP goals and objectives are accomplished through Forest Service (FS) cooperation with State partners, Federal agencies, local units of government, forest landowners and other partners. The FLP identifies and protects environmentally important private forestlands that are threatened by conversion to nonforest uses and provides the opportunity for continuation of traditional forest uses, such forest management activities and outdoor recreation.

The guidelines are organized in three parts:

PART 1-General Programme Guidelines: Programme direction applicable to all aspects of the FLP.

PART 2-State Grant Programme Guidelines: Programme direction applicable to States and Forest Service (FS) Regions/Area/IITF where a State has elected the State grant option and where ownership of lands or interests in lands is vested in a State or subdivision of a State.

PART 3-Federal Acquisition Programme Guidelines: Programme direction applicable to States and FS Regions/Area/IITF selecting the Federal acquisition and ownership process, where ownership of lands or interests in lands is vested in the United States (U.S.).

GENERAL PROGRAMME GUIDELINES

Authority and Purpose of the Forest Legacy Programme (FLP)

Authority

The Cooperative Forestry Assistance Act (CFAA) of 1978, as amended, (16 U.S.C. 2101 et. seq.) provides authority for the U.S. Secretary of Agriculture (Secretary) to provide financial, technical, educational, and related assistance to States, communities, and private forest landowners. Section 1217 of Title XII of the Food, Agriculture, Conservation and Trade Act of 1990 (P.L. 101-624:104 stat.3359; 16 U.S.C. 2103c), also referred to as the 1990 Farm Bill, amended the CFAA and directs the Secretary to establish the FLP to protect environmentally important forest areas that are threatened by conversion to nonforest uses. This authority continues indefinitely. Through the 1996 Farm Bill (Federal Agricultural Improvement and Reform Act of 1996; P.L. 104-127; Title III-Conservation; Subtitle G-Forestry; Section 374, Optional State Grants for Forest Legacy Programme), the Secretary is authorized, at the request of a participating State, to make a grant to the State to carry out the FLP in that State, including the acquisition by the State of lands and interests in lands.

Purpose of the Forest Legacy Programme

The purpose of the FLP is to ascertain and protect environmentally important forest areas that are threatened by conversion to nonforest uses.

FLP seeks to promote forestland protection and other conservation opportunities. Such purposes shall include the protection of important scenic, cultural, fish, wildlife and recreational resources, riparian areas and other ecological values. Traditional forest uses, including timber management, as well as hunting, fishing, hiking, and similar recreational uses are consistent with purposes of the FLP. Both purchased and donated lands and interests in lands through the use of conservation easements and fee-simple purchase are used to acquire forested land meeting Forest Legacy purposes from willing sellers or donors.

Delegations of Authority

The Secretary has delegated authority to administer all aspects of the FLP to the Under Secretary for Natural Resources and Environment (7 CFR 2.20(a)(2)(xvi)) who in turn has delegated the authority to the Chief of the Forest Service (7 CFR 2.60(a)(16). Delegations only apply within the U.S. Department of Agriculture and its agencies. The role of State and Regional programmes, and the right of States to elect the State Grant Option, are contained in the authorizing statute and these programme implementation guidelines.

Description of Terms and Abbreviations

Assessment of Need (AON) is a document produced by a State, or a federally recognized Indian Tribe, in consultation with the State Forest Stewardship Coordinating Committee (SFSCC). The AON contains the an assessment of the forests and forest uses, a description of forces that are converting forests to nonforest uses, describes Eligibility Criteria developed by the State to identify important forest areas to be proposed as Forest Legacy Areas (FLA), and acts as a guide to implementation of FLP in the State.

Assessment of Need (AON) Amendment is a document produced by a State to amend their AON, to add or delete Forest Legacy Areas (FLA), or to modify the Eligibility Criteria. CFAA is the Cooperative Forestry Assistance Act of 1978, P.L. 95-313, 92 stat. 365, 16 U.S.C., 2101 et seq. (as amended through P.L. 107-195, June 16, 2002).

Conservation Easement is a legal agreement a property owner makes with a governmental entity or a nonprofit organization to restrict activities allowed on the land in order to protect specified conservation values. Easement restrictions are tailored to the particular property and to the interests of the individual landowner. All FLP conservation easements are held in perpetuity.

Eligibility Criteria are a set of factors developed by the State lead agency, in consultation with the State Forest Stewardship Coordinating Committee (SFSCC), to evaluate geographic areas to determine if they contain significant environmental values to be considered an 'important forest area' and contain "threats" of conversion to be eligible as a Forest Legacy Area (FLA).

Federal Appraisal Standards are those standards contained in the publication entitled "Uniform Appraisal Standards for Federal Land Acquisitions: Interagency Land Acquisition Conference, 2000," or subsequent amendments or updates.

Forest Legacy Area (FLA) is a geographic area with important forest and environmental values, that satisfies identified Eligibility Criteria and has been delineated, described, and mapped in a State's AON for the FLP. Acquisition of lands and interests in lands for the FLP can only occur within approved FLAs.

Forest Legacy Area (FLA) Boundary Adjustment is a minor change to an existing FLA to create a more logical or manageable boundary.

Forest Legacy Programme (FLP) Project is an individual or series of land or interest in land acquisition transaction(s). The transaction(s) can be on an individual tract or multiple tracts in a distinct geographical area. A FLP project relates to a single funding event in a given fiscal year. FLP projects can have a single parcel that can be completed at one closing or more than one parcel that can be completed in a succession of closings. If a successive FLP project is proposed on a parcel or in a distinct geographic area each transaction is treated as an independent unit in the project selection process and funding is not guaranteed.

Forest Service (FS) is the United States Department of Agriculture Forest Service. Forest Service Region/Area/IITF refers to the field units of the Forest Service responsible for FLP management and oversight within the Forest Service Regions, Northeastern Area (Area) or International Institute of Tropical Forestry (IITF).

Forest Stewardship Plans, or multi-resource management plans, are prepared with the purpose of achieving long-term stewardship of forestland. Such plans identify landowner objectives and describe actions to protect and manage soil, water, range, aesthetic quality, recreation, timber, and fish and wildlife resources, and other conservation values identified on the tract. Plans are to be prepared by a professional resource manager. A Forest Stewardship Plan that meets the requirements of the Forest Stewardship Programme or a multi-resource management plan is required for FLP qualification. The State Forester or equivalent, or their designee must approve the plan.

Full Fee Purchase is a land conveyance where a purchaser acquires all rights, title and interest in a property from a seller or owner. It is also known as fee simple or fee acquisition. Geographic Regions are the collection of States that makeup the National Association of State Foresters (NASF) Regions. The three regions are: North (consisting of the States within the FS Northeastern Area), South (consisting of all the States within the FS Southern Region, and the Territories of the International Institute of Tropical Forestry), and the West (consisting of all the States within the FS Northern, Rocky Mountain, Intermountain, Southwestern, Pacific Southwest (including the Commonwealth of the Northern Mariana Islands, Guam and American Samoa), Pacific Northwest and Alaska Regions.

Indirect costs relate to costs of the management and administration of the FLP. Indirect costs, unlike salary, which is a direct cost, are defined as costs not readily assignable to a specific legacy acquisition.

In-kind contributions are non-cash contributions, including third-party contributions. In-kind contributions must be expenses necessary to accomplish programme activities, and allowable if the Federal Government were required to pay for them. Interests in Land are a right, claim, or legal share in real property that are less than the full title. Land Trust is a nonprofit organization, as described in 501(c) of the Internal Revenue Code of 1986, that protects land by working with landowners who wish to donate or sell fee title or conservation easements to maintain conservation values associated with the land.

Market Value is the amount in cash, or in terms reasonably equivalent to cash, for which in all probability the property would be sold by a knowledgeable owner willing but not obligated to sell to a knowledgeable purchaser who desires but is not obligated to buy. (Uniform Appraisal Standards for Federal Land Acquisitions: Interagency Land Acquisition Conference, 2000, p.4.) Multi-State Entity is a government-established organization involving two or more States or Indian tribes whose jurisdiction encompasses all or portions of the land area of an FLA(s). National Association of State Foresters (NASF) is the organization representing State forestry organizations in all 50 States, the territories and the District of Columbia.

Nonfederal Cost Share refers the nonfederal cost-share required to receive FLP funding. There are three main categories of activities that meet this requirement:

- The value of land, or interests in land, dedicated to the FLP that is not paid for by the Federal government,
- Nonfederal costs associated with programme implementation, and
- Other nonfederal costs associated with a grant or other agreement which meets FLP purposes. The nonfederal cost-share must be documented, and in the case of a grant, must meet the timing, terms and conditions of the grant.

Nonforest Uses

Noncompatible-nonforest uses are uses of the land inconsistent with maintaining forest cover including, but not limited to, activities that result in extensive surface disturbance such as residential subdivisions, commercial development, and mining.

These uses generally should be excluded from FLP conservation easements or land purchases. FLP funds should only be used on parcels with forestland as defined in the State's AON. Compatible-nonforest uses are nonforest uses of the land that may be compatible with forest uses as part of an undeveloped landscape, including cultivated farmland, pasture, grassland, shrubland, open water, and wetlands. These nonforest uses should be less than 25 percent of the total area. Forest Legacy funds should only be used on parcels with forestland as defined in a State's AON. Other funding sources may be used to protect nonforested areas on those parcels with less than the minimum required forest cover.

Nontrust Allotment Lands are privately owned fee simple lands owned by tribal members and if they are forested, are eligible for the FLP when they are located within an approved FLA. Trust lands and reservations are already protected through the trust relationship between the U.S. Department of the Interior and the tribe and are ineligible for the FLP.

Pass-through describes a land transaction whereby a third party, such as a land trust, acquires interests in lands with the intent to convey such interests to a unit of government. The transaction can include a full or partial donation, or sale at market value. Payment in Lieu of Taxes (PILT) is made by tax-exempt entities, including the Federal government, to compensate local jurisdictions for tax revenues foregone as a result of ownership by a tax-exempt owner. Any FLP tract acquired in fee and held by the FS is eligible for PILT payments (entitlement land as defined at 31 U.S.C. 6901). Federal funds for PILT are not authorized for any land or interests in land held by nonfederal entities, or for conservation easements held by the United States.

Programme Funds are FLP funds that are appropriated by Congress and allocated by the FS to three categories: Project funds, Administrations funds, and AON Preparation funds. Project Evaluation Criteria are developed by the States, in consultation with the State Forest Stewardship Coordinating Committees (SFSCC), to evaluate the eligible tracts submitted by interested landowners for inclusion in the FLP.

Relocation refers to the provision in the Uniform Relocation Assistance and Real Estate Property Acquisition Policies Act of 1970 (PL 91-646 or 42 U.S.C. 4601) which requires Federal agencies and programmes to pay for the relocation of a person displaced by a federally funded real estate transaction.

Reserved Areas are designated areas where nonforest uses (*e.g.* house, barn, remote recreation camps, etc.) are or will be allowed, but are inseparable from the land holding and do not have a detrimental effect on the conservation easement values. These areas shall be defined and described in the conservation easement and may be restricted in terms of their use, or provisions made through cost and time to cure and treatment. To the extent possible these areas of non-compliance should be excluded from the FLP project.

Reserved Interest Deed is where the grantee (government) acquires all rights, titles, and interests in a property, except those rights, titles, and interests that may run with the land that are expressly reserved by a grantor (landowner).

Secretary is the U.S. Secretary of Agriculture.

State refers to any of the 50 States, Puerto Rico, Guam, the United States Virgin Islands, the Commonwealth of the Mariana Islands, and American Samoa participating in the FLP. State Forest Stewardship Coordinating Committees (SFSCC) are defined, and their duties are described, in Section 19(b) of the CFAA (16 U.S.C. 2113).

They are chaired and administered by the State Foresters, or equivalent State officials, with membership composed of representatives from the following agencies, organizations, or individuals: Forest Service; Natural Resources Conservation Service; Farm Services Agency; Cooperative State Research, Education, and Extension Service; local government; consulting foresters; environmental organizations; forest products industry; forest land owners; land trusts; conservation organizations; the State fish and wildlife agency; and others determined appropriate by the Secretary. The SFSCC makes recommendations to the State lead agency regarding the AON, AON amendments, and the determination of project priorities. State Lead Agency is that unit of State government responsible for coordinating the establishment and implementation of the FLP in the State, as designated by the Governor or pursuant to State law. The State lead agency is usually a forestry agency, but may be another natural resource agency.

Tribal Assessments of Need-An AON is developed by a federally recognized Indian Tribe in cooperation with the State and the SFSCC. Only nontrust allotment lands are eligible for FLP. Lands or interests in lands purchased under a Tribal FLP can be through a grant to a cooperating State or through the Federal acquisition option.

Working Lands Conservation Committee is a committee of the NASF having coordination and consultation responsibilities within that organization regarding the FLP.

National Environmental Policy Act (NEPA)

NEPA applies to certain proposed actions of the Federal Government. NEPA does not apply to the independent actions of States or private property owners. It has no applicability to a private property owner's use or development of his/her property rights, nor the development of a State's FLP. It could apply to Federal agency actions undertaken on private property if the U.S. acquired a right to permit or deny certain land uses and then proposed to exercise that right, but in such an instance it would be the U.S. that would be required to satisfy NEPA requirements, not the private owner. It should be known that:

- A Programmatic Environmental Assessment and a Finding of No Significant Impact was completed for the national FLP and signed by the Chief of the Forest Service.
- Under the Federal acquisition option, the FS NEPA regulations (Forest Service's Environmental Policy and Procedures Handbook 1909.15-92.1, effective 9/21/92), the acquisition of an individual Forest Legacy tract and/or easement may be categorically excluded from the preparation of an Environmental Impact Statement or an Environmental Assessment unless scoping indicates extraordinary circumstances exist.

POLICIES AND PROGRAMMES RELATED TO URBAN FORESTRY DEVELOPMENT AND MANAGEMENT

There are several policies, programmes and projects issued and implemented for the past several decades which is an indication of continuing concern on the deterioration of urban environment. It is noticeable that new policies and programmes evolve whenever there is a change in administration (a common phenomenon in the Philippines) indicative of lack of continuity of previous initiatives.

The major policies and programmes related to urban forestry are chronologically listed below:

- PD 1153 of Pres. Marcos dated 1976 (Tree Planting Decree to support PROFEM)
  - Requires all able-bodied Filipinos 10 years old and above to plant a tree per month for 5 consecutive years.
  - Certificates of planting and survival— requirement for graduation from school, renewal of job appointment and business permit and approval of retirement from service.
  - "Halamanan ng Bayan" launched by MHS to support this programme. It required each city or municipality to put up a nursery, garden and park.
  - Repealed by EO 287 dated July 25, 1987 because of dictatorial provisions and harsh penalty.
- PD 953 of Pres. Marcos dated July 6, 1976 (Greening of Private Lands Including Residential Subdivisions)
  - Requires private landowners to plant trees extending at least 5 m on each side of the rivers/ creeks.
  - Developers or owners of residential subdivisions and commercial/ industrial lots to set aside 30% of total area as open spaces for parks and recreational areas.
  - Penalizes unauthorized cutting, destruction or injury inflicted on naturally-growing or planted trees or vegetations in any public places.
- LOI 1312 of Pres. Marcos dated April 23,1983 (Establishment and Development of Local Government Forest or Tree Parks Throughout the Philippines).
  - Requires each barangay, municipality or city to establish and maintain at least one forest or tree park of considerable size.
  - MNR (now DENR) to allocate public lands for this purpose and to provide technical assistance and seedlings needed.
  - MHS to ensure that establishment of forest or tree parks is included in the land use plan of each barangay, municipality or city.
  - MILG (now DILG) to appropriate funds and implement establishment and maintenance activities.
- Memo Order Nos. 198 and 199 of Pres. Aquino dated November 9, 1988 (Luntiang Kamaynilaan Programme (LKP)/ Hardin ng Bayan Programme).
  - Issued to help insure healthy environment in Metro Manila (MM) and to serve as model programme for other cities/municipalities.
  - Anchored on the "Hardin ng Bayan" concept wherein each city

or municipality should have gardens or parks of their own, transforming MM into a garden metropolis with lush vegetations, cool and fresh air like the countryside.
  - Objective- to plant 2 million trees in 2-3 years and achieve a desired 1:4 tree-man ratio.
  - For efficient, effective coordinated implementation, an Inter-Agency Committee (IAC) was formed: Co-chair- DENR and MMA (now MMDA); members- DPWH, DOTC, Metro Police Force, DOT, OPS and PMS.
- Memo Cir. No. 5 of Pres. Ramos dated August 27, 1992 (Clean and Green Programme).
  - Similar to LKP (same IAC composition except MMDA as chair/ lead agency) but wider in scope (not only greening but also cleaning activities)
  - Objective- massive planting (0.5 million trees/year or 2.5 million trees in 5 years from 1993-1997) to achieve the ratio of one tree for every 4 persons.
  - Although focused in MM, CGP has nationwide coverage and encouraging cities and municipalities to join nationwide contest for cleanest and greenest city or town.
- EO No. 113 of Pres. Ramos dated July 22, 1993 (Multi-sectoral Tree Planting Activities in Support of ENR Programmes/ECOREV).
  - Scope/ Objective – regreening and rehabilitation of all open and denuded lands of public domain, idle lands, private lands and other suitable areas (both urban and rural) including rehabilitations of coastal and marine areas.
  - DENR to identify, assess and designate suitable area for planting and management and to provide technical assistance to participating agencies.
  - LGUs implement the programme in their respective level and set up counterpart funds.
  - Private sector participation encouraged via MOA or other appropriate arrangements with DENR.
- EO No. 118 of Pres. Ramos dated August 12, 1993 (Mandating the active participation of all government agencies nationwide in urban greening through an Adopt-A-Street/Park Programme)
  - Objective – greening of streets and parks in urban centers.
  - Requires all government offices and government owned/controlled corporations to adopt a street or park in coordination with concerned LGUs, NGOs and private sector by planting appropriate species and maintaining them for at least 5 years using their own funds/resources. and other resources.

- DENR to manage and coordinate the programme through a designated National Coordinator.
- Project to be turned over to concerned LGU for maintenance and protection.

- DENR-DILG-DPWH-CSC Joint Memorandum Circular No. 1 dated December 17,1993 (Implementing Guidelines for EO 118-Adopt-A-Stree/Park Programme)
  - Described the roles of each participating agency and outlined the schemes in the identification, selection and adoption of a street or park to be developed.
  - DENR to provide assistance to "adopters" in selecting suitable streets or park sites, in providing necessary planting materials and in monitoring performance.
- OPLAN SAGIP PUNO Programme of FMS-NCR/DENR launched on June 5, 2000.
  - Conceived as a component of "Lets Go Green Programme" of former DENR Secretary Antonio Cerilles.
  - Application of appropriate silvicultural treatments (*e.g.* removal of nails, wires/cables, water sprouts; surgical treatment of injured stem or root) to prolong life span and promote good health and vigor of trees planted in parks and along thoroughfares and streets in MM.
  - Supplemented by public awareness campaign.
  - DENR enters into MOA with participating agencies (*e.g.* subdivision homeowners association, city/ municipal government, NGOs, etc.)
  - DENR's role — conduct inventory and assessment of damaged/ injured trees; undertake appropriate silvicultural treatments; conduct information dissemination and training on tree care and maintenance; provide technical assistance and planting materials to sustain the project.
  - LGU's role — provide tree care and maintenance crews to sustain the project; assist DENR in information dissemination on maintenance and protection of trees.
- Proclamation No. 396 of Pres. Arroyo dated June 2, 2003 (Enjoining the active participation of all government agencies including government-owned or controlled corporations, private sector, schools, civil society and citizenry in tree planting activity and declaring June 25,2003 as Philippine Arbor Day).
  - Objectives- to promote multi-sectoral participation in tree planting nationwide; to develop greater awareness on the importance of trees in environment, health and human life.

- Participating agencies, LGUs, schools, etc. to identify areas to be planted in coordination with agencies which have jurisdiction over such areas *e.g.* DENR in case of public lands, LGUs in areas within their jurisdiction, DND for military lands reservation, DOT for ecotourism areas, etc.
- DENR, LGUs and schools — to establish and maintain nurseries.
- Respective participating agency/ instrumentality — to maintain and protect the planted seedlings.
- DENR — to provide technical assistance to all participants.

In general, the following goals and objectives are common to the Urban Forestry (UF) policies and programmes described above:

- to provide/maintain green, clean and beautiful environment;
- to promote public awareness on the importance of trees (promote environmental consciousness);
- enhance people's participation in the programme;
- promote multi-sectoral collaboration, cooperation and support; and
- in the case of LKP and CGP, the specific objective is to attain a 1:4 tree to person ratio to sustain ecological balance.

As strategy to enhance successful implementation of the project, DENR is usually tasked to provide technical assistance in planting, site and species selection, and maintenance operations, including provision of the planting stocks. Understandably, the DENR is also looked up to as the lead agency when inter-agency collaboration is involved in the programme. On the other hand, the city, municipal and barangay governments, which have jurisdiction over the project site, are usually tasked to maintain and protect the tree parks established and streets planted. They are also required to provide counterpart funds and other resources needed for these projects.

At the end of each programme, there seems to be no serious post – project accounting or evaluation of outputs and accomplishments, including evaluation of success and failures. This may be attributed to the fast rate of turn – over of urban forestry/greening programmes being implemented. Another reason maybe lack of manpower and resources to monitor all the projects. For instance, in the case of Metro Manila, the Urban Forestry and Law Enforcement Division Office of FMS – NCR/DENR only has a small unit (Cooperative Planting Unit) under the Urban Forestry Section which is tasked to do the monitoring activities. Needless to say, the synthesis of lessons learned is an important input for planning and formulation of new programmes (*i.e.* we do not have to " reinvent the wheel" so to speak).

GREEN SPACE DEVELOPMENT AND MANAGEMENT

To develop, implement and coordinate the urban forestry programmes for Metro Manila, DENR created the Urban Forestry Division (UFD) in 1988, now

Urban "Forestry and Law Enforcement Division (UFLED), under the Forest Management Services (FMS) of DENR – NCR. The Urban Forestry Section under the UFLED has 2 units: the Planting Stock Unit, which is responsible in planting stock production and distribution, and the Cooperative Planting Unit which is tasked to monitor collaborative UF programmes like LKP, CGP and Adopt – A – Street/Park Programmes. The Urban Forestry Section is also currently implementing the OPLAN SAGIP PUNO programme. The urban forestry/greening programmes of DENR- NCR was conceived and adopted to make Metro Manila into a green metropolis. Among its activities include production of planting stocks; establishment and maintenance of mini – forests; greening of main thoroughfares, side streets and islands; establishment of joint nurseries with LGUs, NGOs, POs, schools; providing technical assistance and training on proper site preparation, choice of species and proper planting and maintenance; harnessing the cooperation and involvement of the public via information campaign; periodic assessment of greening activities; processing and issuance of balling permits, and implementation of Oplan Sagip Puno programme.

At the city or municipal government level, specific urban forestry/greening offices were created under the Mayor's office. Some of these offices are ad – hoc in nature while the others are permanently institutionalized in the city government structure. The greening offices of all the 6 cities have a mandate on green space development and management although in 3 cities (Makati, Pasig, Mandaluyong), they also included cleaning, waste management and pollution monitoring and control as their other mandates. In addition to the greening office, each city also has either a committee or Tasks Force created to enhance active participation of other sectors in the greening and cleaning activities.

The capabilities of these greening offices in terms of manpower, available facilities and financial resources were also assessed by Palijon (2000). He found that: a) majority of these offices felt the need for additional technical staff to fully implement their greening programmes such as horticulturist, forester/arborist, and landscape architect; b) the greening offices have inadequate facilities, equipment and tools needed for their programmes; and c) all cities, except Makati, have insufficient and unsustainable budget for their greening programmes. In the case of the other mega-cities, the DENR regional office does not have specific urban forestry unit unlike in the NCR. Their urban forestry programme activities are being handled by the Reforestation Section under the Forest Resources Development Division of the Forest Management Services, except in Cagayan de Oro where a focal Programme Unit for Urban Forestry was recently created by the RED attaching such unit directly to the RTD-FMS. At the city government level, the City ENRO, Clean & Green Office, Task Force Clean & Green and Cleaning & Beautification Committee are usually the offices or bodies attached or created under the Mayor's Office responsible for management of urban forestry programmes just like those of the NCR.

PHYSICAL ACCOMPLISHMENTS OF URBAN FORESTRY PROGRAMME

The total number of seedlings planted in the urban greening programme by city/municipality in Metro Manila from 1988 to 2002. The 17 cities and municipalities comprising Metro Manila (NCR) have an aggregate total of 2,212,488 seedlings planted. The highest number of seedlings planted was in Quezon City (0.72 million seedlings or 34% of total), followed by Manila City (0.31 million seedlings or 14% of total). On the other hand, the total population of Metro Manila in 2000 was 9.93 M people.

Based on the statistics, the estimated tree to person ratio is 1:6 or 1:9 assuming 80% or 50% survival of the seedlings planted, respectively. These ratios are short of the 1:4 target ratio set by MPFD. However, the estimated number of trees is most likely an underestimation because the trees already existing prior to 1988 were not accounted or tallied. For the other mega-cities, the attainment of 1:4 ratio can not be ascertained due to absence of data on the number of seedlings planted on different years and the absence of inventory data on the number of trees already existing in these areas.

Accomplishment for Mini Forests and Parks

The minimum number of mini-forests or parks targeted by the MPFD for Metro Manila was 60 for the 8.2 million residents and this was based on the assumption that there should be 1 mini-forest or park per 100,000 – 150,000 residents. Since the present population of Metro Manila (NCR) is about 10M, there should be at least 67 mini-forests or parks already established and maintained.

As of 1994, there were already 472 parks established in different parts of Metro Manila, which is more than enough (*i.e.* 7 times more) compared to the target ideal number. Almost 50% of the total number of parks in Metro Manila were established in Quezon City. Between 1996 to 2001, eight more mini-forests were established in Metro Manila covering about 18.20 hectares. Hence, for the entire Metro Manila, a total of 480 mini-forests or parks have already been established as of 2001.

Compound Planting

As envisaged in the MPFD, at least 570 compounds or grounds should have been planted or landscaped by year 2000 in all six priority cities. Since there is no breakdown per city, the target maybe equally allocated at 95 compounds or grounds per city. There is no available information on the number of compounds or grounds planted or landscaped in the different mega-cities.

Nevertheless, for Metro Manila, data is available on the number of seedlings planted. For the past 14 years, about 1.1 million seedlings have been planted on different types of compounds mostly in the government offices and schools in Metro Manila.

Greenbelt Development

For Metro Manila, the MPFD targeted 100 km of roadsides planted or greenbelt developed from 1990 – 2000. Since there is also no data available on actual length of roads planted or greenbelt developed, we can use the available data on number of seedlings planted along thoroughfares and streets for extrapolation. There were 673,813 seedlings planted along major thoroughfares and barangay roads of Metro Manila for the past 14 years. Due to limited space along road shoulders, it is assumed that there will be a single line planting on each side of the road. At 2m spacing, there will be 1,000 seedlings required per km of road. Assuming these 673,813 seedlings planted included 50% replanting, it is estimated that about 337 km of roads and thoroughfares should have been planted in Metro Manila. This is 3.4 times more than the target of MPFD.

For Iloilo City, the DENR-FMS Region 6 reported 34 adoptors (government agencies, universities, private companies, NGOs, etc.) involved in Adopt-A-Street/Park Programme since 1994. Of the 34 adoptors, 31 adopted streets spanning 17 km while 3 adopted perimeters of parks covering about 1 km. Since the MPFD's roadside planting target from 1996 to 2000 is 35 km, only about 50% of the target was accomplished.

In the case of Cagayan de Oro City, there were 10.5 km roadside planting done in 1995 to 1999 in barangays Balubal and Indahag and along Lumbia Airport Road (DENR-FMS Region 10). On the other hand, the City ENRO conducted 8 km highway tree planting/urban greening in barangays Cugma, Tablon, Bugo and other urban barangays (City ENRO – Cagayan de Oro City). Thus, the DENR and City ENRO's roadside planting efforts fall short (*i.e.* only 46% accomplishment) of the MPFD's 1996-2000 target of 40 km. However, for the year 2001-2005, the MPFD's target of 20 km greenbelt development was already accomplished through the tree and bamboo planting project undertaken in 2002 along 20 km riverbanks in barangays Balulang, Lumbia, Bayanga, Mambuaya and Dansolihon through a contract awarded by the ENRO to MARBEMCO, a people's organization (City ENRO - Cagayan de Oro, 2002).

URBAN FORESTRY OR GREEN SPACE DEVELOPMENT AND MANAGEMENT

Numerous issues/problems/constraints were raised during the 1993 Policy Seminar – Workshop on Participatory Urban Forestry Development for Environmental Enhancement in Metro Manila, and those presented by Palijon (1998, 2000) based on his analysis of green space management strategies in

Metro Manila. On the national level, several issues/constraints/problems were listed based on the 1993 Regional Annual Reports on Urban forestry by the DENR regional offices. Among the issues identified are as follows:

- Institutional
  - Absence of urban forestry section in DENR regional offices except NCR
  - Poor or unsustained inter–agency coordination
  - Lack of organizational capability
  - Absence of Master Plan on urban forestry development
- Social and political
  - Vandalism/indifference or apathy by the general public
  - Uncontrolled squatting/ encroachment
  - Improper use of street corridors
  - Damage or removal of trees due to road expansion, installation of electric and telephone lines, construction of new buildings, etc.
  - Land use conflicts/lack of proper zoning
  - Lack of tree protection ordinances or laxity in implementation of existing laws in some cities of Metro Manila.
  - Ineffective urban population and migration control programmes
- Biophysical and technical
  - Lack of water during summer
  - Frequent occurrence of typhoons
  - Unsuitable soil physical and chemical properties
  - Poor species-site compatibility
  - Incompatible design of drainage systems and other underground utilities which restrict growth of planted trees.
  - Lack of maintenance & protection and monitoring of growth/ survival performance of trees planted
  - No M&E system for urban forestry programme.

NATURAL FOREST PROTECTION PROGRAMME ON RURAL HOUSEHOLD LIVELIHOODS

Assessing the impacts of ecological restoration programmes on the livelihoods of local residents, is imperative for two main reasons. First, such programmes have severe equity impacts and these should be understood and mitigated when designing such policies. High biodiversity areas are often also home to poor, rural communities, while the benefits of conservation may go to wealthier residents of the same, or other, countries. In developed countries, conflicts between conservation policy and land use also occur, but those deriving income from the land are much more likely to have secure property rights to that land than households in developing countries who may be affected by similar policies. The second reason for accounting for the impacts of conservation programmes on local residents is that a failure to do so may lead to conflict over the resources, reducing the potential effectiveness in environmental terms. In this chapter, we estimate the impact on local household livelihoods of the Natural Forest Protection Programme (NFPP), which is the largest logging ban programme in the world that aims to protect watershed and conserve natural forests.

Existing studies that have attempted to evaluate the impacts of this programme have concluded that while it has been effective in reducing logging and increasing reforestation, these benefits have come at a significant cost in terms of the impacts on household livelihoods. In this chapter, we focus on evaluating the impacts of the NFPP on two important facets of the livelihoods in the affected areas, namely those of household income and employment opportunities. We argue that it is important to understand these particular impacts as the NFPP is part of a process of more general Chinese forest policy reform which involves a shift from focussing simply on timber output, to new objectives of: i) enhancing the role of forests in ecological rehabilitation and environmental protection; ii) increasing timber supply through commercial investment; and iii) promoting rural wellbeing and poverty reduction through agro-forestry. This means that even if the NFPP is contributing to the first objective, it cannot be considered successful if it undermines objectives ii) and iii) through negative impacts on household incomes and on their incentives to invest in future forest activities.

Measuring the impacts of the NFPP on household incomes and labour opportunities is not a straightforward matter, mainly because the Chinese economy has been undergoing huge changes during the period that the programme has been in place. Further, we cannot observe household income and labour decisions both in the presence and the absence of the programme, and therefore face an identification problem.

To address this, we treat the NFPP as a natural experiment, using panel survey data to compare the changes in income and labour opportunities over time in the areas where the programme was in place with changes in the areas where it was not introduced. In order to ensure the robustness of our results, we estimate the changes in income resulting from the programme using various parametric and non-parametric policy evaluation technique and we find that the NFPP has had a negative impact on incomes from timber harvesting. However, it has actually had a positive impact on total household incomes from all sources.

Further, as the increase in total income appears to be driven by an increase in income from off-farm employment, we also estimate the impact of the NFPP on household participation in off-farm labour markets. We predict that the reduction in the marginal return to labour in forest activities resulting from the ban on logging would lead to a reallocation of labour from forestry to off-farm occupations. Using similar evaluation methods as for the income impacts, we estimate the effect on off-farm employment, and find that it has increased more rapidly in NFPP areas than non-NFPP areas. This result is strongest for employment outside the village. We note that the increase in total household income may not correspond to an increase in household welfare because if that were the case, non-NFPP households would be expected to participate in the off-farm labour market to a greater extent as well.

BACKGROUND TO THE NFPP

The Natural Forest Protection Programme (NFPP) was introduced by the Chinese government in response to serious drought in the Yellow River area in 1997 and flooding of the Yangtze River in 1998. Both of these events were understood to have been brought about by poor agricultural practices and the removal of natural forests. The Sloping Land Conversion Programme (SLCP) was intended to address the former problem, while the NFPP addressed the latter. The NFPP applies to 17 provinces and autonomous regions in the upper reaches of the Yangtze River and the upper and middle reaches of the Yellow River. These provinces contain 73 million hectares of natural forests, which amount to 69% of the total natural forest area in China.

Restore of Natural Forests

The objectives of the programme are: to restore natural forests in ecologically sensitive areas, protecting and enhancing biodiversity; to plant forests for soil and water protection; to increase timber production in forest plantations to meet national demands for timber and contribute to economic development of rural areas; and to protect existing natural forests from excessive cutting. These objectives are being pursued with the use of, firstly, a ban on logging in natural forests across many parts of the country; and secondly, measures to encourage the development of new plantation forests which should increase timber supplies while reducing pressure on natural forests. These activities have been supported with funds from the Chinese government, directed specifically at: i) afforestation and forest protection, including mountain closure, tree planting, construction of sapling bases, payments for forest tending and forest fire prevention; ii) compensation for unemployed state forest workers, retirement pensions for state forest staff and some compensation to local governments for losses of forest taxation revenues.

Existing Studies

The NFPP is the largest of its kind in the world, and therefore there is considerable interest in its impacts. The most significant study so far has been carried out by Xu et al (2002) and found that while there have been positive impacts on forest cover, there have been certain negative effects such as losses in employment in the state forest sector; reductions in local government finances; and losses of employment and income for households dependent on the state and collective forest sectors. Other studies have largely been case studies of forest management by households within a small number of villages or townships, for example Shen (2001), Weyerhauser et al (2006), and Demurger and Fournier (2003).

General Findings

The NFPP is largely accepted to have reduced timber harvesting in natural forest areas. As it has been implemented in important watersheds of the major rivers of China, and in areas of high biodiversity, this is expected to have positive environmental impacts, in particular on soil erosion and water conservation. However, two key environmental issues have been identified: the first is that there are suggestions from case studies that the programme has reduced the incentives to actively manage forest land by monitoring against illegal logging and forest fires and by thinning and maintaining the timber stands, particularly where the ban has been applied to village plantations. The second issue is that the reduction in timber supply within China is expected to have negative effects on neighbouring countries. This raises concerns because much of the imported timber originates in Asia-Pacific countries where unsustainable harvest practices and illegal logging are common.

Much of the research into the impacts of the NFPP has focussed on the impact on the state forest sector as this was initially targeted in the programme. During the early stages, it was estimated that around 1 million employees would have to be laid off. Many of these have been re-employed in forest stewardship, while those who were not re-employed were offered compensation of three times their average salary for the previous year. While there are some case in which compensation has not been paid in full, and where state enterprises are unable to pay full pensions for retired workers, greater losses in state forest areas have been experienced by those who were not formally employed by the state sector. There have also been impacts on the provision of public services by local governments as their revenues have dropped significantly in areas where forestry played an important role in the local economy. For example, in some counties in Sichuan Province, timber revenues represent more than 80% of total income from taxation; as timber production falls, these revenues will fall accordingly.

Economic Benefits

Yang (2001) also highlights economic benefits from the NFPP. He argues that the 1998 floods affected hundreds of millions of people and caused extensive damage, and that the result of this was major expenditures on flood defences against future disasters, including 7.8 billion Yuan on embankments along the

Yangtse. The financial and social costs of both the floods themselves and of protection against future events are real costs that should be included in an evaluation of the programme. A further benefit he suggests is diversification of economic activities, giving the example of Aba Prefecture which has promoted tourism as an alternative to logging. The prefecture experienced increases of around 130% in both numbers of tourists and tourist revenues between 1997 and 1998 and now receives 30% of GDP from tourism.

FINDINGS RELATING TO COLLECTIVE FORESTS

The NFPP was primarily targeted at state forest areas, but in many places it has been extended to cover collective forests as well. A number of studies have argued that where households and communities are prevented from accessing collective forest land, this infringes their existing rights to use that land, and their ownership of the trees on the land.

There is little in the way of quantified economic impacts on collective forest communities. Katsigris (2002) looks just at Sichuan Province, and finds that timber output from collective forests had fallen to 6% of previous levels by 2000, resulting in revenue losses of approximately 1 billion Yuan for the whole sector. She also found a 65% reduction in the number of forestry related township enterprises, with a corresponding fall in employment and output of 53% and 30% respectively. The same study also finds losses in tax revenues and increases in employment, and reductions in overall economic growth rates in some counties with collective forest areas, although those with alternative development opportunities were observed to recover quickly from 2000 onwards.

Chen *et al* (2001) measure income changes for three townships in Li County, Sichuan Province. In one of the townships, they observe a 17.6% reduction in per capita income between 1998 and 2000. In the other two townships, per capita incomes fall by around 1/3. These reductions are explained by losses of employment in the timber harvesting or processing sectors, or loss of revenues from provision of services to temporary migrants who were previously working in the forest sector. Shen (2001) studies two villages in Aba Prefecture, Sichuan Province and finds that households have not only lost income from the timber sector, but also through loss of access to non-timber forest products, including fuelwood, and access to grazing land for livestock. In general, the NFPP does not prohibit the collection of non-timber products, but in some cases is has either been disallowed by the local government or access has been prevented by the closure of transport routes to forest areas.

Contribution of this Study

Despite the number of studies that have been carried out into the impacts of the NFPP, there has been an absence of analysis using quantitative information to assess these impacts, as those cited have relied on qualitative case study information. Case studies such as those carried out for the Xu *et al* (2002) study; and by Shen (2001), Weyerhauser et al (2006), and Demurger and Fournier (2003), were each focused on a small number of villages, and based on interviews with key members of state forest enterprises and forest communities. Their conclusions have raised considerable concerns about the NFPP but have as yet not been investigated more widely. Therefore, this chapter attempts to evaluate the impacts of the programme, and revisit some of the conclusions of the existing studies, using a survey of 285 households in 40 villages in the south of Guizhou Province.

The existing case studies are largely based in the southwestern region of Sichuan and Yunnan Provinces. So far there has been no research into the impacts in the southern collective forest areas. As the forest sector in the southwest is rather different in character to the forest areas in the south of China, data from Guizhou will be the only assessment of the impacts of the ban in the southern provinces. Finally, the existing studies of the NFPP are based on information that was collected shortly after the logging ban was introduced, so this chapter supplements those findings by using data from 2005, when some adjustment to the ban will have taken place and the medium term impacts are more apparent.

CATEGORIES OF PROGRAMME FUND IN FOREST POLICY

Forest Legacy funds are allocated to one of three categories: Project Funds, Administration Funds, and AON Preparation Funds. FLP funds may not be used for monitoring and enforcement.

PROJECT FUNDS

Project funds are those used to directly purchase lands or interests in land joining the FLP. Project funds may be expended by the State lead agency or the FS, as applicable, to cover transaction costs, including but not limited to: appraisals and appraisal review, land surveys, closing costs, establishing baseline information, title work, purchase of title insurance, conservation easement drafting, and other real estate transaction expenses for those tracts. Project funds may also be expended to facilitate donations of land or interests in lands to a qualified and willing donee for FLP purposes, by paying for expenses directly related to the donation, including but not limited to, land surveys, conservation easement drafting, title work, and establishing baseline information.

For an outright donation of a conservation easement or land, FLP programme funds may not be used to pay for an appraisal. In the case of a partial donation of a conservation easement or land, an appraisal meeting Federal standards is required to determine the value of property. FLP funds may be used for appraisals on these partial donations. When Federal funds are used to purchase real property, including conservation easements, appraisal and acquisition work procedures must meet Federal standards.

Administration Funds

Administrative funds are the portion of funds used for day-to-day programme management at all levels. Administration funds may be used for a variety of activities, including FLP programme administration, personnel and overhead, and all activities identified as eligible uses of project funds to prepare projects and potential projects. Forest Legacy funds for administration shall be kept to a minimum. As a goal, all attempts should be made to keep administration funds under 15 percent of the total funds appropriated.

AON Preparation Funds

AON preparation funds may be made available to States to help defray the cost of preparing, or amending an AON.

Process for Allocating Funds to Forest Service Regions/Area/IITF

Following passage of the annual appropriations bill, the FS Washington Office develops the Forest Legacy Programme Direction and allocates funds to the FS Regions/Area/IITF for distribution. The allocation process differs for each fund category described below.

Allocation of Project Funds

Allocations to FS Regions/Area/IITF are based on the results of the national project selection process and the appropriations bill. Under the State grant option, FS Regions/Area/IITF will award grants to States for specific, identified projects.

Allocation of Administration Funds

The FS Washington Office distributes administration funds to FS Regions/Area/IITF. Each FS Region/Area/IITF in consultation with the States requests these funds to meet their needs and the needs of the participating states in their Region/Area/IITF. Administration funds are also used by the FS Washington Office to fund programme management functions. Administration funds will be granted to States under the State grant option separately from project funds.

Allocation of AON Preparation Funds

The FS Washington Office distributes AON preparation funds to the States by way of FS Regions/Area/IITF. These funds are requested by FS Regions/Area/IITF to meet the needs of their States to develop new AONs or amendments.

Redirection and Reprogramming of Funds

Due to the nature of real estate transactions, FLP projects may change in scope, cost or fail completely. These changes can result in unspent or excess funds for some projects while others may need additional funding to bring them to completion.

In order to maximize the efficient and effective use of FLP project funding, the FS will either redirect or request reprogramming of funds. Redirection is a shift of funds from one congressionally approved project to one or more other congressionally approved project(s). Reprogramming is a shift of funds that exceeds an increase or decrease of 10% per project not to exceed $500,000 to an existing project, or shifting of any amount of funds to a project not previously approved by Congress.

Regional Redirection Process

FS Regions/Area/IITF may redirect up to an increase or decrease of 10% per project not to exceed $200,000 of project funds that are excess or unspent from one project to one or more other Congressionally approved project(s) within the FS Region/Area/IITF which is underfunded and where there is a substantiated need (*e.g.* loss of other funding sources, appraisal documenting increased cost, etc.) to bring the project to completion. Project funds over $200,000, or those that cannot be redirected by the FS Region/Area/IITF, will be released for the national process.

FS Regions/Area/IITF will notify the FS Washington Office before a redirection takes place and report these actions periodically. All funds from failed projects will be released for the national process.

National Redirection Process

The FS Washington Office, through consultation with FS Regions/Area/IITF, may redirect up to an increase or decrease of 10% per project not to exceed $500,000 of project funds that are excess or unspent from one project to one or more congressionally approved project(s) which is underfunded and where there is a substantiated need (*e.g.* loss of other funding sources, appraisal documenting increased cost, etc.) to bring the project to completion. In addition, when funds have not been spent or contractually obligated within two years of receipt of funds, they revert to the FS Washington Office via the appropriate FS Region/Area/IITF.

The FS Washington Office will:

1. Assess the extent of unspent or returned funds on a periodic basis;
2. Facilitate selection and funding of underfunded projects not addressed by the regional process or between FS Regions/Area/IITF; and
3. Notify Appropriations Subcommittees of any redirection action taken by the FS FS Regions/Area/IITF or FS Washington Office.

FLP Cost Share Requirements

The CFAA directs that, to the extent practicable, the maximum Federal contribution for total programme costs may not exceed 75 percent. To assure programme-wide cost share goals are met, each project budget must include a minimum nonfederal contribution of 25 percent. This nonfederal cost-share must meet Forest Legacy purposes. It may consist of:

- The value of land, or interest in land, dedicated to the FLP that is not paid for by the Federal government;
- Nonfederal costs associated with programme implementation; and
- Other nonfederal costs associated with a grant or other agreement that meets FLP purpose.

The nonfederal cost-share must be documented, and in the case of a grant, must meet the timing, terms, and conditions of the grant. The cost-share can occur at any phase of the FLP including planning, developing future projects, acquisition, capital improvement, management, or administrative activities. When a grant is involved, the cost-share must occur within the life of a grant and meet all grant requirements.

Federal requirements identify the grant period as beginning when the grant is formally awarded and ends after two years to ensure that the federal funds are spent promptly. However, a grant may receive a maximum extension to five years. Allowable costs shall be determined in accordance with the 7 CFR 3016, "Uniform Administrative Requirements for Grants and Cooperative Agreements to State and Local Governments," and any amendments to this regulation. Donations of land or interests in land must be documented to count as part of the nonfederal cost-share.

The title does not need to be transferred to the State or federal government in order for the donation to qualify as cost share. However, if in the future, the donated lands are conveyed or the rights or title are modified in a way that is inconsistent with the purposes of the FLP then the State must restore the cost share value dedicated in the grant agreement.

The value of donations may be included as part of the nonfederal cost-share if all of the following are met:

- The donation contributes to the objectives and priorities of the State FLP as set forth in the AON;
- All or part of the tract being donated must be within the boundaries of an FLA, and may include National Park, National Forest, National Wildlife Refuge, or other Federal land boundary, or within the boundaries of an area designated through an analogous State programme with goals compatible with the FLP and be within an FLA;
- The donor documents their desire that value of the interests may be used as cost share for the FLP project;

- The donation of land or an interest in land must contain perpetual covenants to assure that the tract will be managed in a manner compatible with the goals for which the FLA was established;
- The donee (holder of donated rights) is a unit of government or a non-profit conservation organization (land trust) that meets the eligibility requirements for holding a conservation easement established by the Internal Revenue Service and has as its purpose the management of lands or interests in land consistent with FLP purposes;
- If the donation is in the form of a conservation easement then the deed needs to contain a provision that directs all of the easement holder's proceeds from a subsequent sale or exchange of interests in land be used in a manner consistent with the conservation purposes identified for the subject interests in lands;
- The respective portion of the donation must not have been previously credited towards any Federal program's nonfederal cost share; and
- The State lead agency approves the donation as contributing to the cost-share.

Acquisition of Lands or Interests in Lands

FLP acquisitions may be outright full fee purchases, or acquisition of development rights or other rights conveyed through a conservation easement. Except in the case of a full and complete donation of land or an interest in land, if any Federal funds are used in the acquisition of Forest Legacy tract the following shall apply:

- Federal appraisal standards must be met, including appraisal review by a qualified Review Appraiser;
- The landowner must be informed in writing of the market value and that sale of the property is strictly voluntary;
- The landowner must be notified in writing that the property will NOT be purchased if negotiations do not result in amicable agreement;
- Federal payment to the landowner for lands or interests in lands is not more than the market value as determined by an appraisal meeting Federal appraisal standards;
- The title acquired must be free of encumbrances inconsistent with the purposes of the FLP. Title insurance may be secured for the full value of the encumbered property, but is not an alternative to an acceptable title; and
- If relocation is involved the requirements in the Uniform Relocation Assistance and Real Estate Property Acquisition Policies Act of 1970 (PL 91-646 or 42 U.S.C. 4601) must be followed. The FS will be advised in advance of any acquisition involving relocation.

- In the case of acquisition of interested in lands, the development of a Forest Stewardship Plan or multi-resource management plan that has been approved by the landowner and the State Forester or designee and Baseline Documentation Report shall be prepared prior to project closing.

All FLP acquisitions of lands or interest in lands are perpetual and therefore run with the land. Although any remaining interests held by the landowner may be subsequently conveyed, future owners are still bound by the terms and conditions of the conservation easement. At the same time, future owners shall retain full control of the rights that are not acquired by the FLP, and shall be subject only to those restrictions that the present landowner has conveyed to the Federal, State, or local government. Compatible nonforest land uses (cultivated farmland, pasture, grassland, shrubland, open water, and wetlands) are desirable land uses in many FLAs. FLP funds should not be used for any property not meeting the State's definition of forested land in the AON, unless there is a written plan scheduling reforestation or afforestation. Programmes to conserve farms, ranches and similar land uses may be used in conjunction with the FLP to protect properties where there are mixed forest and compatible nonforest uses.

Conservation easements are required to contain language pertinent to the purpose of the FLP and a reversionary provision to ensure the conservation investment of FLP into the future. During the development of tract specific conservation easements, a determination will be made as to whether the acquisition of mineral rights, prohibition on reserved areas, or an exclusion of the area that does not comply with FLP, would be necessary in order to protect the other rights that are being considered for acquisition. In some situations, it may be impossible to protect environmentally important forest areas pursuant to the purpose of the FLP without acquiring the mineral rights.

The FLP conservation easement holder (Federal, State or local government) is responsible to assure that baseline documentation contains all the information necessary to monitor, manage and enforce the easement. Where the conservation easement is a tax-deductible gift, and the owner retains rights to the property, the Internal Revenue Service (IRS) holds the donor responsible for providing sufficient baseline data "to establish the condition of the property at the time of the gift." However, this does not eliminate the FLP need for baseline documentation.

Baseline documentation describes or depicts a tract of land and its attributes on the day it becomes restricted by an easement. This documentation is required on all FLP tracts and is completed prior to project closing. Documentation of the property should include a map of the area drawn to scale showing all existing man-made improvements or incursions such as roads, buildings, fences or gravel pits; an aerial photograph of the property taken as close to the date the property is restricted as possible; and on-site photographs, especially of significant features. The above should be accompanied by narrative descriptions of tract attributes and other pertinent information.

States and landowners are encouraged to display the official FLP signs on the FLP property using the signs. The posting of FLP tracts helps promote public awareness, recognition and support for the programme. Landowner permission should be secured before posting any signs. Costs associated with sign posting can be covered by FLP project or administration grants or States may use such expenses as FLP cost share. Signs should be inspected during the annual monitoring of the FLP tract and repaired when in poor condition. FLP sign art and programme logos may be used by FLP partners for items that contribute to the purpose of awareness (*e.g.* brochures, workshops, outreach efforts, posters, FLP information packets, web sites etc.)

Multi-State Identification of Forest Legacy Areas

States are encouraged to cooperate in the identification of FLA that cross State or Tribal boundaries and to work together to coordinate acquisitions of lands or interests in lands that have complementary purposes. States may elect to jointly use an existing or new multi-State or regional entity to identify FLAs or develop FLP projects that cross State boundaries. The entity must be a government-established organization, whose jurisdiction encompasses all or portions of the land area of the FLA States involved. However, programme implementation is undertaken by the individual States (State Grant Option) or by the FS (Federal Option). The entity conducting a multi-State identification of FLAs is responsible for:

- Obtaining approval from the appropriate States or Indian tribes for FLAs within their boundaries,
- Cooperating with appropriate SFSCCs,
- Obtaining public comments on the identification of FLAs, and
- Complying with all other requirements of these guidelines.

9

Global Temperature and Climate Change in Forest Ecosystems

Global temperature and climate are largely determined by the balance of incoming energy from the sun, minus outgoing radiation. Incoming light radiation from the sun has short-wavelengths and can readily pass through the atmosphere, but after being absorbed and re-radiated from Earth's surfaces the out-going infra-red radiation has longer wave-lengths and is less able to pass through the atmosphere. The so-called "greenhouse gases" absorb and re-radiate a portion of the outgoing long-wave radiation back towards earth, acting like a heat-trapping blanket. Even slight changes in the ratio of incoming and outgoing solar energy have significant influence on our global climate system. Even though greenhouse gasses make up less than 1 per cent of Earth's atmosphere, our global climate is quite sensitive to changes in their concentration.

Ice-core data from Greenland and Antarctica tells us that atmospheric levels of carbon dioxide ($CO_{2)}$ vary somewhat predictably with cycles of ice ages and warm inter-glacial periods. The ice cores also show that atmospheric CO_2 is increasing almost 100 times faster today than during past climate cycles, and that current concentrations of CO_2 are higher than at any time in at least the last 800,000 years. Given the difficulty of rapidly changing our resource-intensive lifestyles, we'll be lucky if global atmospheric CO_2 concentration merely doubles. More likely it will go much higher before we control our appetite for fossil fuels and land exploitation.

While CO_2 is of primary concern among greenhouse gasses, there are others such as methane (CH_4) that contribute to global warming. CO_2 is unique in that is has a very long, approximately 100 year, "residence time" in the atmosphere.

Concentrations of CO_2 in the atmosphere will likely remain far above "normal" for centuries, because the millions of tons of CO_2 released to the atmosphere during the agricultural revolution, the industrial revolution, and the automobile revolution will not reach a new equilibrium until biological and geophysical processes (in the oceans and on land) have a chance to capture and store most of the "extra" carbon. We have a moral obligation to leave future generations with choices and opportunities for survival. We must avoid irreversible harm to the planet's life support systems including a livable climate and function ecosystems that sustain life.

ATMOSPHERIC CONDITIONS

The climate of any given place is determined chiefly by its latitude, its elevation and meteorological forces that operate over wide areas and at high as well as low altitudes. But within the broad limits established by these major factors, local modifications of the "microclimate" are effected by topography and forest cover. Although these effects vary greatly with different combinations of slope, aspect, kind and size of trees, density of stand and area covered, certain generalizations as to their character are possible.

WIND MOVEMENT

The first thing that one notices on entering a forest on a windy day is the relative calm. So great is the difference that one seems to have stepped suddenly into another world. The mechanical obstruction offered by the trees deflects upward a large part of the moving mass of air and slows down the velocity of that which enters the forest. Dense stands with heavy foliage naturally have the greatest effect. Thus, a forest of deciduous broadleaf trees has more effect on wind velocity in summer than in winter.

Observations indicate that the velocity of the wind in the interior of a forest may range from about a tenth to two thirds of that in adjacent open country, depending on the character and condition of the forest. The stronger the wind, the greater the reduction of actual velocity, but not of relative velocity. Under normal climatic conditions wind velocities within a forest are typically low, seldom averaging more than a few miles per hour.

This influence on wind movement is felt not only inside the forest itself, but for a considerable distance to the leeward. Shelter belts consisting of several rows of trees are often planted to take advantages of this fact, particularly in regions where lack of natural forest cover gives the wind an unbroken sweep over long distances. Intensive studies in the plains region of the United States show that shelter belts may have a measurable influence on wind velocity to a distance of as much as fifty times the height of the trees. The greatest effect usually comes within a distance of from three to five times the height of the shelter belt, where the velocity of the wind in its lee may be 20 to 30 per cent of that in the open.

This reduction of wind movement brought about by forests and shelter belts adds to the comfort of man and beast and increases the production of crops grown to the leeward of their sheltering influence.

HUMIDITY

In general, the air in a forest is somewhat moister than that in the open, although the difference in absolute moisture content is less than would be expected in view of the large amount of water transpired by trees. Relative humidity, which varies with temperature as well as with the absolute amount of water vapour in the air, may be as much as 11 per cent higher in the forest. As in the case of temperature, the influence of the forest is much more pronounced on maximum and minimum humidities than on mean humidities.

PRECIPITATION

No subject in the field of forest influences has aroused more controversy than the effect of the forest on precipitation. Whether that effect is negligible or significant has often been argued with more heat than light. Popular belief in the efficacy of forests as "rainmakers" was formerly much more widespread than it is today. Passage of the Timber Culture Act of 1873 was undoubtedly facilitated by the hope that tree planting in the prairie region would increase the supply of water, as well as of wood. Unfortunately, so many complicating factors are involved as to make positive proof one way or the other extremely difficult.

The argument that forests increase rainfall and other forms of precipitation is based chiefly on the fact that they transpire enormous quantities of water. Trees not only use more water in the process of growth than does any other form of vegetation, but may even add more moisture to the atmosphere than is evaporated from an equal area of open water. In effect they act during the growing season as a pump, moving water from the soil to the air in almost unbelievable quantities. And whatever influence transpired water may have in increasing precipitation is strengthened by the lower temperature in and above forests, by the fact that they add to the effective height of the land and by the friction which they offer to wind movement.

All these factors, it is asserted, tend to result in more frequent and greater precipitation-an influence that may be regional as well as local. Raphael Zon, for example, in his report to the National Waterways Commission argued that "forests in broad continental valleys enrich with moisture the prevailing air currents that pass over them and thus enable larger quantities of moisture to penetrate into the interior of the continent. The destruction of such forests... affects the climate, not necessarily of the locality where the forests are destroyed, but of the drier regions into which the air currents flow."

The opposing argument is that the major forces controlling precipitation operate over such wide areas — continental or even intercontinental in extent — that the presence or absence of forests can have virtually no effect.

Furthermore, precipitation depends not only on the amount of moisture in the air, but on conditions favorable to its release. How forests affect these conditions, if at all, is uncertain. Many attempts have been made to measure the influence of forests on precipitation, with widely varying results and with even more diverse interpretations of their meaning. Although most of these studies show significantly greater precipitation in the forest than in the open, the difference may be more apparent than real. This is primarily because the greater wind velocity in the open reduces the amount of precipitation caught by rain gauges. There are also other complicating factors that make it difficult to obtain strictly comparable measurements.

The opinions expressed by Joseph Kittredge in his book *Forest Influences* probably represent the point of view held by most scientists today.

He concludes that forests have no appreciable effect on cyclonic precipitation, but that they may increase local rainfall in temperate climates by not more than 3 per cent and that rainfall in clearings in the forest may be about 1 per cent more than in similar situations in the open.

Evaporation: Forests tend to decrease the amount of water evaporated from the soil under their immediate cover and to some distance in their lee. This is due to their influence on wind movement, temperature and humidity. One of the major objectives in the planting of windbreaks or shelter belts is to reduce evaporation and thereby increase crop production in the fields that benefit from this protection, which, commonly extends over a distance many times the height of the windbreak.

TEMPERATURE AND RADIATION

Forests exercise a moderating influence on air temperature as well as on wind movement. Here again the extent of the influence is greatest in dense stands with heavy foliage, which have the maximum effect on incoming and outgoing radiation. Protection from solar radiation, which under extreme conditions may be in the forest only 1 per cent of what it is in the open, reduces maximum temperatures throughout the year. In the forests of the United States the reduction in monthly maxima may range from about 3° F. in January to 8° F. in July.

Similarly the slowing down of outward radiation from the earth generally raises minimum temperatures throughout the year. The effect of a forest on monthly minima is less than on monthly maxima, both in summer and winter, but may amount to more than 6° F. in the United States.

Since maximum temperatures are normally lower and minimum temperatures higher in the forest, the net effect of a forest cover is to reduce materially the range in temperature at all times of the year and also to reduce slightly the mean temperature. Evergreen trees have more influence than those with deciduous foliage. Although the actual figures may seem small, it must be remembered that the effect on absolute maximum and minimum temperatures is much greater than the effect on mean temperatures; the moderating influence of the forest is most noticeable and most welcome on extremely hot or extremely cold days.

SOIL TEMPERATURES AND CONDITIONS

TEMPERATURE

Soil temperatures are affected by a forest cover in the same way as air temperatures but to a much greater degree. Mean daily maximum temperatures at the soil surface may be reduced in summer by as much as 50° F. and absolute maxima even more. The influence of the forest varies inversely as the distance from the surface, but under some conditions may be recognizable to a depth of 30 feet.

Minimum temperatures show a similar but less pronounced increase. This influence is particularly noticeable in the winter, when soils usually freeze later and less deeply in the forest than in the open. Sometimes the soil in the forest remains unfrozen while that in adjacent open fields is frozen to a considerable depth. On the other hand, in situations where there is a heavy blanket of snow in the open and a very light blanket under a dense forest canopy, this relationship may be reversed.

These influences are due partly to the reduction of insolation and radiation by the overhead forest cover and partly to the insulating effect of the litter and humus of the forest floor. These reduce maximum temperatures more than they increase minimum temperatures and hence the net effect is a cooling one.

In both cases the effect is beneficial. During the summer young plants are protected from high temperatures which in exposed situations in the open may be so extreme as to kill the living tissue; while during the winter reduction in the depth and period of freezing diminishes surface runoff by permitting the penetration of more water into forest soils.

MOISTURE CONTENT

Forests tend to decrease the amount of moisture in the soil by intercepting precipitation, by retaining water in the forest floor and by transpiration. They tend to increase the amount by reducing surface runoff, by increasing the permeability of the soil and by decreasing evaporation from it.

Much precipitation in a forest never reaches the ground because it is first intercepted by and then evaporated from, the leaves, twigs, branches and trunks. This loss may run as high as one tenth of an inch per shower. The amount of water that actually reaches the ground may vary from none in light showers to the great bulk of the total precipitation in heavy rains. Interception naturally increases with the density of the stand and of the foliage and is much greater with deciduous trees in summer than in winter. When the precipitation reaches the ground it is first absorbed by the forest floor, which has a field moisture capacity (that is, an ability to retain water against the pull of gravity) of from one to five times its own dry weight and when saturated contains much larger quantities. Its absorptive and storage ability is much greater than that of mineral soil, the field capacity of which usually ranges from 10 to 50 per cent of its dry weight. The thicker and the more decomposed the litter and humus of the forest floor, the more water it will hold.

The amount of water that actually gets through to the mineral soil may vary from zero to 100 per cent, depending on the amount of precipitation and on the condition of the forest floor. With light showers in times of drought all of the precipitation may be retained by the forest floor; while with heavy showers and in wet periods little or none may be so retained. On land without vegetative cover and particularly with heavy soils, the surface layer tends to become hard and impermeable, with marked reduction in the amount of precipitation that sinks into the main body of the soil. In the forest, on the other hand, the litter slows down surface runoff and thus gives more time for the water to be absorbed by the humus, which acts as a huge sponge. More important still, the humus prevents muddying of the underlying mineral soil, which remains porous and thus absorbs and retains much more water than does exposed soil in the open.

Enormous quantities of water are removed from the soil by transpiration. A well-stocked forest of mature trees uses more water than does any other form of vegetation. It may require from 100 to 1,400 pounds of water for every pound of dry matter produced. Transpiration is far greater in summer than in winter and during the growing season is somewhat greater with broadleaf deciduous trees than with conifers. Its tendency is obviously to reduce the moisture content of soil in the forest as compared to that in the open. On the other hand, the reduction of evaporation in the forest, which was mentioned in the discussion of atmospheric conditions, tends to maintain the moisture content at a higher level than in the open.

The net effect of these conflicting influences varies widely with the character of the topography, the soil, the climate and the forest. So far as any generalization is possible, it may be concluded that soils under a forest cover tend to be somewhat drier than similar exposed soils in the open on fairly level ground and somewhat moister on steep slopes where more water is lost by surface runoff in the open than by interception and transpiration in the forest.

Erosion and Stream Flow

Of all the influences of the forest on its environment, the most obvious and the least controversial is the reduction of surface runoff of water. This prevents or substantially checks soil erosion and modifies stream flow.

COMPOSITION AND STRUCTURE

Forests have a marked influence on the composition and structure of the underlying soil. Every year they add to the forest floor large quantities of leaves, twigs and branches, which under normal conditions are constantly decomposing to form humus. Part of this humus gradually mixes with the mineral soil beneath and also supplies it with soluble compounds carried downward by percolating water.

The amount of organic material in the forest floor varies widely with differences in climate, soil and vegetation. Recorded data run from less than 2 tons per acre in old-growth longleaf pine in Florida to nearly 120 tons in a forest of birch, sugar maple and spruce in New Hampshire. In general, high temperatures result in its rapid decomposition and disappearance. Ordinarily there is a gradual transition from undecomposed litter through partially decomposed duff and humus to a mixture of organic and inorganic material and eventually to mineral soil. An exception occurs in coniferous forests in the north where slow decomposition of the needles results in acid "raw humus." This raw humus does not mix well with the underlying mineral soil and in extreme cases may be peeled off like a blanket.

The usual effect of a forest cover is to improve both the chemical and the physical characteristics of the soil. Nitrogen, calcium, phosphorus, potassium and other elements are made available in larger quantities. Even more important is the action of organic materials in making the soil more friable and crumbly. They tend to make light soils, such as sands, heavier; and heavy soils, such as clays, lighter. These effects are strengthened by the constant growth and death of the tree roots. While growing they keep pushing into and loosening up new areas of soil; when dead they add organic material to the soil and leave channels through which water may percolate more readily. Porosity of frozen soil is also increased in the forest, where it is more permeable than in the open.

ACCELERATED EROSION

Soil is constantly being moved by water and gravity from higher to lower elevations. Under normal conditions, with an undisturbed vegetative cover, this is a slow process known as "geologic" erosion. Over the centuries it has built up fertile alluvial soils in the valley bottoms. With the removal or disturbance of the vegetative cover, a striking change takes place. Larger and larger particles of soil and even huge boulders, can now be moved and the rate of removal is greatly increased.

The beneficial process of geologic erosion is replaced by the destructive process of "accelerated" erosion. Among the harmful results are deterioration or ruin of the lands where the erosion takes place and where the coarse detritus is deposited, siltation of reservoirs, impairment of the quality of water for municipal and industrial uses, destruction of fish habitats, clogging of river channels and increase in the volume of floods. All kinds of vegetation serve to check erosion, but by far the most effective are well-stocked forests and well-sodded grasslands. Their influence is primarily due to their ability to reduce the amount and velocity of surface runoff.

The amount of material that water can carry at a given velocity is directly proportional to its volume. As the velocity increases, however, there is a much more than proportional increase in the cutting and carrying power of a stream. Thus, if the velocity of water is doubled, its cutting power is increased fourfold, its carrying power thirty-two-fold and the size of the material it can carry sixty-four-fold. These facts explain the occurrence of "mud-rock" flows, the transportation by small streams of boulders weighing many tons and up to a six-thousand-fold increase in the rate of erosion following destruction of the forest and the forest floor.

During ordinary storms much, if not all, of the precipitation is absorbed by the forest litter and humus, retained a short time and then passed on gradually to the mineral soil beneath. The humus layer is not comparable, as has sometimes been alleged, to a blotter on a slate roof, the efficacy of which in storing water ceases as soon as it becomes saturated. Except in "raw humus," there is no sharp line between the organic surface and the inorganic subsurface material in the forest.

On the contrary, there is a gradual transition from the top layer of newly fallen, undecomposed litter to partially and completely decomposed humus, to a decreasing mixture of organic with inorganic material and finally to mineral soil. Thus there is a constant and uninterrupted downward movement of water from the temporary storage provided by the upper layers of the forest floor into the larger and more permanent soil reservoir beneath. From there the water moves slowly as a subsurface flow to springs, streams and lakes, from which it is in time evaporated and again precipitated in the course of the hydrologic cycle.

Tree roots also help to keep the soil from being washed away in gullies and along the banks of streams and tend to check the occurrence of landslides. Erosion by wind is virtually nonexistent in the interior of a forest and is greatly reduced for some distance to its lee. In many ways forests are influential in preventing the development of accelerated erosion — the greatest threat to the world's soil resources. Fortunately their influence is most effective where the danger is most acute, as on steep slopes, in clayey soils and with heavy rainfall. Wind erosion can also be prevented or greatly reduced by properly placed windbreaks in open country, as in the Great Plains of the United States.

FLOODS

What effect these influences of the forest have on floods is a hotly debated question. Certainly forests cannot prevent an abnormally high flow of streams when heavy precipitation falls on saturated or frozen soils. If the storage basin is already full, there is nothing for any additional supply of water to do but to flow off by way of the nearest stream. On the other hand, when the soil reservoir is well below the saturation point, there may be far less surface runoff from forested areas than from those not covered by vegetation, even in severe storms.

By and large there can be no doubt that forests tend to reduce flood flow in small watersheds in hilly country. One would expect them to have a similar influence on large rivers made up of the flow of many small streams. Here, however, the situation is complicated by many factors, the combined effect of which it is difficult to evaluate. For example, the way in which precipitation is distributed over a river basin as a whole and the rate of flow of the tributary streams largely determine whether their peak discharges reach the main river simultaneously or in successive waves.

Obviously the first combination will result in a much higher flood than the second, even with the same or perhaps even a smaller total runoff. Therefore, the clear-cut influence of the forest on the flow of streams from small watersheds at the headwaters of large rivers may not always be reflected in the flow of the main river.

Nevertheless, the general tendency will be for the forest to decrease both maximum and total runoff and to increase minimum runoff, throughout the river basin. The action of the forest in reducing erosion also has an important influence on flood flow and flood damage.

Small streams with steep slopes issuing from deforested or burned-over watersheds may carry several times as much solid material as water. One instance is on record in southern California where solids comprised 88 per cent of the total flow. As the stream flattens out, the larger and heavier materials are, of course, deposited; but even rivers with small gradients may carry heavy loads of sediment. This material adds significantly to the volume of the stream and increases its erosive power, particularly in time of flood. It may, in fact, be even more important than the volume of water as a source of damage.

DISTRIBUTION OF RUNOFF

For all practical purposes the total stream flow from any watershed consists of the precipitation less the sum of the losses from interception, transpiration and evaporation. To the extent that the forest increases interception and transpiration, it tends to reduce stream flow; and to the extent that it decreases evaporation, it tends to increase stream flow. The net effect depends on the relative weight of these influences. It will obviously not be the same under all environments or at all seasons of the year.

Measurements from forested and from deforested or partially forested areas in Switzerland, Colorado and North Carolina have shown a strong tendency toward smaller annual runoff from the forested areas. The difference was particularly marked in the spring, when melting snow and heavy rains caused the maximum seasonal runoff. The greater discharge of water from the nonforested areas also carried with it much more eroded material.

The total flow of a stream is made up of two main parts: the surface runoff and the subsurface or ground-water runoff. The forest has a profound effect in reducing surface runoff and thereby increasing the amount of water available for subsurface runoff. The evidence is conclusive that in most regions surface runoff is very small or negligible from areas of undisturbed vegetation, while it may amount to half of the precipitation where the vegetation has been destroyed or seriously disturbed. However, the outstanding influence of forests in this respect is greatly weakened when the forest floor is destroyed by repeated fires, even though the trees themselves are not killed.

Reduction of surface runoff results in more uniform flow of a stream throughout the year. Maximum flow is nearly always less and minimum flow usually more from forested than from comparable nonforested areas. The tendency is to avoid sharp peaks of flow in the spring and deep troughs in the summer. The total amount of flow may be less, but its greater uniformity is highly desirable.

WATERSHED MANAGEMENT

From the point of view of water supply, the ideal management of the forest, as a part of the broader field of watershed management, is that which will produce the maximum total runoff, well distributed throughout the year, with minimum erosion.

Such management may not always be identical with that which will produce the largest supply of wood. Serious conflicts will, however, be the exception rather than the rule. When they do occur, the method of management to be selected must be based on relative values, among which water for domestic and industrial use, irrigation, power, navigation and recreation may rank high. Striking evidence of the basic importance of conserving both soil and water resources is afforded by the greatly increased attention being paid by such agencies as the Soil Conservation Service, the Tennessee Valley Authority and the Forest Service in the United States to the promotion of more effective watershed management, in which forest planting and improved forest practices play a prominent part.

Wood, the Universal Raw Material

No less prominent than the influence of the forest on climate and runoff is its influence on industry. In fact, its tangible products affect our economic activities even more clearly and more directly than do the intangible services that it renders. They provide opportunity for the profitable employment of labour and capital in many important industries and furnish a wide variety of indispensable consumers' goods. Wood is sometimes known as "the universal raw material" because it is so widely distributed and can be used for so many purposes. Although for some uses it has been largely displaced by other materials, there are few articles into the composition, manufacture, or transportation of which it does not enter. While perhaps no more "indispensable" in modern civilization than many other materials, such as iron, cement and glass, its range of use is certainly much broader than most of them.

For thousands of years, until the advent of coal, oil and gas, wood constituted man's chief source of fuel. Even today more than half of the total consumption of wood in the world goes for this purpose. Another third in the form of lumber and structural timbers is used for the construction of buildings of all kinds. Except in urban centers and where forests are scarce or lacking, most people continue to live in wooden houses.

Lumber and dimension stock are in turn remanufactured into innumerable articles such as toothpicks, toys, sporting goods, musical instruments, handles, boats, caskets and furniture. In the United States approximately an equal amount of lumber (about one seventh of the total cut) goes into boxes, crates and dunnage used in the shipment of commodities. Large additional amounts of wood are used for railroad ties, poles, piling, fence posts, mine timbers, excelsior, shingles, cooperage, veneer and plywood. Modern adhesives are greatly helping to expand the use of wood in the fields of ply and laminated construction and are making practicable the salvage of much low-grade material. Advances in engineering techniques now make possible the use of wood in members requiring high strength and long spans, where metal was formerly regarded as indispensable. Even more spectacular advances have taken place in the chemical utilization of wood. The demand for paper and paper products seems unlimited. Each year larger and larger quantities of wood go into the manufacture of newspaper, book and magazine paper, writing paper, wrapping paper, wallpaper, building paper, pasteboard cartons and paper board. From "dissolving pulp" come cellophane, rayon, artificial wool and many plastics. Today one could, if one chose, be clothed wholly in textiles that originated in the forest.

A lively imagination is required to visualize the many other chemical substances that can be derived from wood. Among these are wood gas, potash, charcoal, acetone, methyl alcohol, sugar, ethyl alcohol, yeast, salicylic acid and vanillin. Potentially the forest is the source of large quantities of motor fuel and food.

During World War II, when gasoline was scarce, many automobiles in Europe were operated entirely on wood gas, which is still used to a considerable extent in trucking. Another potential substitute for gasoline is ethyl (grain) alcohol produced by the fermentation of sugar, which in turn has been produced by the hydrolysis of wood cellulose. The alcohol can also be converted into lubricating oil and synthetic rubber. Wood sugars have already been used extensively in Europe as feed for livestock and are a possible source of food for human beings. They can also be turned into yeast of high nutritive value by inoculating them with the appropriate organisms.

Edible carbohydrates and proteins are among the innumerable products obtainable from wood cellulose, which in its original state is a highly indigestible product. Lignin, the other main constituent of wood, is a comparatively unknown substance, of which little use is now made but which is likely in time to provide a wide variety of important chemical products.

Many substances other than wood and its derivatives come from trees and are ready for use with relatively little further processing. Among these are nuts, maple syrup, chewing gum, tannin, dyes, rubber and resin. The last yields turpentine, universally used as a solvent in paints and varnishes and in the production of synthetic camphor; rosin, used by violin players and baseball pitchers and in the manufacture of soap, varnish and paper; pine oil, used in the flotation process of separating metals from their ores and in the textile industry in the fixation of dyes; and pitch, for which chemists are just beginning to find profitable uses.

Many pages would be required merely to list the useful substances that come from trees. Wood, once regarded by some as obsolescent, is now used more widely and for more purposes than ever before. Its inherently desirable properties of high strength in relation to weight, its workability, elasticity, nonconductivity of heat and electricity and beauty can be enhanced and its undesirable properties of shrinking and swelling, inflammability and susceptibility to attack by insects and fungi can be mitigated by methods developed by modern technology. Wood as wood, properly treated and handled, is one of the most useful materials known to man; while cellulose, lignin and resin are a veritable treasure house of chemical products already in use and yet to be discovered. The significance of forest products is emphasized further by the facts that forests are both renewable and more efficient as producers of solid substance (ligno-cellulose) than is any other crop. Here is at least one case where we can have our cake and a rich cake at that and eat it too. Industrially the age of the forest is ahead of us as well as behind us.

Forest and Wood-Using Industries

This situation makes it possible for man, through intelligent handling of the forest resources, to raise his standard of living by the continued and increasing production of a host of valuable, often indispensable, raw materials and finished commodities. It also provides widespread opportunities for the profitable employment of land, labour and capital.

CARBON MOVE IN AND OUT OF THE ATMOSPHERE

There is a fixed amount of carbon on planet earth which is distributed among several carbon reservoirs or pools in the atmosphere, biosphere, hydrosphere, and lithosphere. In the grand scheme, carbon is neither created nor destroyed but continually moves between these various pools owing to the operation of natural and human-induced processes. The root cause of global climate change is that human activity has shifted massive quantities of carbon to the atmosphere from forests, soil, and fossil deposits.

In the atmosphere carbon is stored as CO_2, methane (CH_4), and other organic compounds. Carbon moves *into* the atmosphere from decomposition of organic matter, respiration by living organisms, combustion, volcanic activity, burning fossil fuels, degassing of waterbodies, etc. Carbon moves *out of* the atmosphere via photosynthesis, rock weathering, dissolution in water, etc. All plants, including forests and many micro-organisms, use photosynthesis which takes CO_2 out of the air to build sugars that can be used by the cell to build cellulose or other complex carbon molecules that comprise plant biomass.

This process is called "primary production" and it feeds the bottom of the global food chain. Virtually all life on earth, including humans, relies directly or indirectly on photosynthesis. Most terrestrial plants share a significant portion of their photosynthate with soil organisms, a cooperative relationship that builds a large and complex underground ecosystem that also stores carbon. Plants shed dead leaves and wood which also builds carbon stores in the soil.

In the hydrosphere (e.g. the oceans) carbon is stored mostly as dissolved $CO_2$and other dissolved organic compounds that originated in some photosynthetic life form. Carbon moves *into* the ocean from the atmosphere and biosphere via dissolving of gaseous CO_2 in cold seas, leaching from soil, and input of organic matter from river systems and the biosphere. Carbon moves *out of* the ocean primarily via photosynthesis (*e.g.* phytoplankton and cyanobacteria), degassing of warm seas, and deposition in marine sediments.

In the biosphere carbon is stored as live or recently dead plants, animals, and micro-organisms both in the ocean and on land (*e.g.*, forests and soils). Forests dominate the terrestrial carbon cycle, harbouring 86 per cent of the planet's above-ground carbon and 73 per cent of the planet's soil carbon. Carbon enters *into* the biomass pool via photosynthesis, then becomes entrained and cycled through the entire global food chain. Carbon moves *out of* the biomass pool through decomposition and respiration or through deposition in long-term storage in soil or geologic and fossil deposits.

In fossil deposits, the carbon from long-dead plants and animals are stored as coal, oil, "natural gas," or kerogen. These can be thought of as both "ancient sunlight" and "ancient atmosphere." Carbon moves *into* the fossil pool via deposition and storage in low-oxygen conditions. Carbon moves *out of* fossil pool mostly via industrial exploitation and combustion.

In the non-fossil lithosphere carbon is stored in carbonate rocks such as limestone and chalk. Carbon moves *into* these geologic structures mostly through ocean deposition. A portion of the oceanic carbon is taken up to make the shells of marine organisms that fall to the deep ocean floor where they may be subducted beneath the earth's crust and end up in long-term geologic storage, *e.g.* the Cliffs of Dover. Carbon moves *out of* the lithosphere mostly via volcanic activity and human industry such as the manufacture of cement which heats limestone and releases significant quantities of CO_2.

The advent and diversity of life on earth has had a profound impact on the global carbon cycle and now plays a fundamental role in determining whether or not we have a livable climate. The abiotic carbon cycle that existed before the proliferation of life was less stable than the carbon cycle that developed after marine organisms started to make calcium carbonate shells and deposit carbon in deep storage which has helped buffer CO_2 extremes over long time scales. Scientists have found a correlation between biodiversity and levels of atmospheric CO_2 over the last 370 million years.

Human activity, mostly in just the recent era, has dramatically reallocated global carbon stores from the other carbon reservoirs into the atmosphere where it can influence our climate. For example, burning fossil fuels and heating limestone to make cement move carbon from long-term fossil and geologic storage into the atmosphere. Logging kills trees - stops carbon-uptake via photosynthesis, and moves carbon from living forests and soil into the atmosphere. Land uses such as agriculture, livestock grazing, and draining swamps move carbon from the soil to the atmosphere.

Climate Change Affect the Pacific Northwest

While predicting the local *weather* is an uncertain science, *climate* prediction is actually more accurate because the focus is on large-scale trends rather than local details. We know that the planet as a whole is almost certain to become warmer on average, and scientists expect an acceleration of the hydrologic cycle as warmer temperatures lead to increased evaporation from the oceans and more transpiration from plants. However, the effects of climate change will not be uniform around the globe. Significant uncertainty remains about how global trends will express themselves regionally. Future climate in the Pacific Northwest is even more uncertain because of complex topography and uncertain changes in precipitation, but our close proximity to the moderating influence of the Pacific Ocean likely offers a slight buffer from climate extremes.

The Pacific Northwest should expect continued climate variability. Existing cycles of cool-wet winters and warm-dry summers will likely continue, though they will be superimposed on a warmer average climate. Both floods and droughts have been part of our past and will almost certainly be part of our future, and both will likely get worse, but we don't know if these climate extremes will be expressed with more frequency or more intensity, or both.

It is reasonable to expect more precipitation, mostly during our existing wet seasons. More of our winter precipitation will fall as rain instead of snow, so storage of water in snowpacks will likely decrease (on average). We should expect milder winters, earlier melting of snow packs, earlier spring run-off, longer periods of summer low stream flow, and more drought.

Importantly, earth's biogeochemical systems are complex and not at equilibrium. There are many feedbacks that lead to non-linear behaviour, so we should NOT expect climate changes to be slow and predictable. Small changes in CO_2 and global temperature can lead to large and/or rapid changes in climate and ecosystems. Accordingly, the rate of current and future global changes may be unprecedented, chaotic, and highly disruptive.

CLIMATE CHANGE COULD AFFECT ENTIRE FOREST ECOSYSTEMS

Thousands of years ago, in cooler and wetter times, Bishop pine trees are thought to have proliferated along the West Coast of the U.S. and Mexico. Now, stratus clouds — the low-altitude clouds known locally as "June gloom" — help keep the trees growing on Santa Cruz Island, Santa Rosa Island, and on one island off Baja California. Other than these locations, Bishop pine trees grow only farther north in California where it is cooler and wetter. Mariah S. Carbone, first author of a new paper, titled "Cloud Shading and Fog Drip Influence the Metabolism of a Coastal Pine Ecosystem," and her co-authors, studied the influence of clouds on the largest Bishop pine forest of Santa Cruz Island. Carbone is a postdoctoral fellow with UC Santa Barbara's National Center for Ecological Analysis and Synthesis (NCEAS). Their study was published in the journal *Global Change Biology*.

"When people think about climate change, they're often thinking about temperature and precipitation," said Carbone. "When you think about precipitation, it's rain and snow, depending on where you are. What this study showed is that you can have really important water inputs coming from clouds that influence the carbon cycle." She explained that clouds are one of the largest uncertainties in global climate, and that the forest ecosystem interactions with clouds have a large impact — one that has rarely been studied before, particularly in Southern California.

Changes in cloudiness or cloud height with global warming will alter the types of forest ecosystems that grow in coastal California, explained Carbone. While predictions about how rainfall will change in California over the next 50 years are uncertain, fog and low-stratus clouds might decline as sea-surface temperatures warm. Like the relict Bishop pines on Santa Cruz Island, and the majestic coastal Redwood forests in Northern California, forests remove and store carbon from the atmosphere, and thus are important for understanding greenhouse gas concentrations in the future. These coastal forests are also highly valued for their aesthetic and recreation qualities. Fog is often present in coastal California during May through August. It occurs when the land warms in the spring and summer, and moist ocean air is pulled over cold, upwelling coastal waters. This moist air then condenses under a stable atmospheric inversion layer, creating low clouds or fog banks.

"The finding that summer fog strongly impacts carbon cycling highlights the need for improved understanding of whether we should expect coastal summer cloud behaviour to change in a warmer world," said second author A. Park Williams, a former graduate student in UCSB's Geography Department, now at Los Alamos National Laboratory. "A change in summer fogginess could produce temperature, moisture, and carbon feedbacks in coastal ecosystems that easily swamp out the effects expected from increased greenhouse gases alone," said Williams. The carbon cycle involves the movement of billions of tons of carbon between the oceans, lands, and atmosphere every year. Increased amounts of carbon dioxide, a potent greenhouse gas, ¬¬are injected into the atmosphere and contribute to climate change.

The scientists received help from NASA and NOAA in collecting information. "We used satellite data to map the spatial distribution of cloud cover across the western portion of Santa Cruz Island," said senior author Christopher J. Still, formerly an associate professor at UC Santa Barbara, now at Oregon State University. "We had some evidence of a cloud cover gradient from the coast to inland, based on our own observations and our weather station data, but the satellite data — from NOAA and NASA satellites — really allowed us to quantify this gradient across the western half of the island and in particular at our two study sites," said Still.

Still said that the authors believe this study is unique in linking clouds to soil microbial activity. "While most previous research on fog and ecosystems has focused on its role in plant-water relations — such as the well-known linkages between fog and coast Redwoods — we show that soil microbial activity and metabolism in these coastal ecosystems may be very dependent on the light but frequent fog drip that occurs during the rainless summer months," said Still. The study is also unique in linking fog and low clouds to the carbon cycle of a coastal ecosystem, including its impact on tree growth and soil respiration. "What we found was that there are two major effects of the clouds," said Carbone. "One is shading, which keeps the temperatures cooler and the soil moisture higher.

The other is fog drip, which is a water input into the soil." The soil moisture content was not depleted as rapidly where the trees were under more cloud cover, and thus they were able to grow longer into the dry season. Water pulses from fog drip immediately stimulated microbes in the litter and soil below, but did not directly enhance pine tree activity. The microbial activity releases carbon dioxide into the atmosphere.

"This study provides us with a greater understanding of how fog and the low coastal clouds which Californians commonly call 'the marine layer' affect plant growth and the activities of microorganisms in the soil," said John Randall, scientist with The Nature Conservancy, which owns and operates Santa Cruz Island in cooperation with the National Park Service and the University of California. "Their findings of the importance of fog and low clouds in influencing plant growth and survival and microbial metabolism in the soil will help us understand how changes in coastal fogs and low clouds that may accompany climate change will affect the forests and soils on Santa Cruz Island, other islands off the coast of California and Mexico, and all along the coast of the adjacent mainland."

The work on Santa Cruz Island, conducted in cooperation with The Nature Conservancy, was funded by grants from the Andrew Mellon Foundation and the Kearney Foundation of Soil Science. The NOAA Climate and Global Change Postdoctoral Fellowship and the National Center for Ecological Analysis and Synthesis have supported Carbone.

PROTECT FORESTS FROM THE PERILS OF CLIMATE CHANGE

Jerry Franklin points out that "forest management can either exacerbate or reduce the effects of climatic change on the productivity and biological diversity of northwest forestscapes." To increase the chances that we will continue to enjoy the diverse benefits we receive from northwest forests, we must maintain and enhance their ability to respond to change. The key components of such a strategy are:

- Maintain biodiversity in all its dimensions. This will be critical, because genetic diversity is like a library of possibilities that have worked well during past climate variability, representing the sum of "tools" available for the future.
- Protect intact native ecosystems where species relations have stood the test of time and remain robust;
- Provide refugia and allow species to migrate. Buffer and expand protected areas to provide connectivity along climatic gradients. Manage the entire landscape to be amenable to dispersal of native species.
- Protect streams. Cold water fish are particularly vulnerable to climate change because of increased winter flooding, reduced summer stream flow, and increased stream temperature. To mitigate expected effects on fish we should provide generous riparian buffers to help shade

streams and maintain lower stream temperatures. To render streams more resilient to hydrologic extremes, such as flooding, we should manage whole watersheds to improve their ability to absorb, store, and slowly release water. This can be accomplished in part by reducing disturbance of vegetation and soils, reducing road densities, and retaining abundant woody debris.

LOGGING RELEASES SIGNIFICANT AMOUNTS OF CARBON

Not surprisingly, logging accelerates the transfer of carbon to the atmosphere by killing trees that would otherwise continue to capture and store carbon through photosynthesis and growth. Killing trees also stops them from pumping carbon into the soil where much of the carbon in forests is stored. Logging actually accelerates the rate of decomposition of wood via several mechanisms. By removing the forest canopy and exposing the soil to more sunlight, logging raises soil temperature which increases the rate of decay. Logging also breaks up woody material in the forest thereby decreasing the average piece size and increasing the surface area exposed to microbial decomposition. Finally, logging debris is often burned on site or as part of an industrial process.

Traditional logging also increases the risk of disturbances. Logging increases wind damage by creating exposed edges and increasing wind speeds within forest stands. Logging often increases the wildfire hazard by making the stand hotter, dryer, and windier; by moving the most flammable small fuels from the forest canopy to the forest floor (*i.e.*, logging slash) where they are more available for combustion; and by initiating the establishment of dense stands of young trees with interlocking branches (resinous fuels) close to the ground. Logging roads also increase the risk of human-caused fire ignitions and spread tree diseases like Port Orford cedar root disease that kill trees and release carbon. Scientists estimate that a large fraction of all the carbon transferred to the atmosphere by humans has been released due to forest exploitation. In recent decades CO_2 emissions resulting from human-induced changes to forests exceed CO_2 emissions from all motor vehicle sources combined, but forest releases are less than total emissions from all uses of fossil fuels. After logging an old-growth forest, the site remains a net source of carbon for more than 20 years, and depending on the conditions, the site does not rebuild pre-logging carbon stores for a century or more. As a result of widespread clearcutting and aggressive slash burning, the Pacific Northwest has contributed huge quantities of carbon to the atmosphere.

Increase Carbon Storage in Forests

Here in the Pacific Northwest we live in the midst of a globally significant carbon pool that should be nurtured and conserved to help keep carbon out of the atmosphere. Temperate old-growth forests of the Pacific Northwest contain some of the highest amounts of biomass per acre measured anywhere in the world. About half of the dry weight of forest biomass is comprised of carbon. The latest IPCC Mitigation Report notes that "Forest-related mitigation activities can considerably reduce emissions from sources and increase CO_2 removals by sinks at low costs..." The IPCC also states that more than 1/3 of the potential mitigation available from forests is located outside the tropics and half of the forest mitigation will come from changes in forest practices, rather than simply preventing deforestation.

The objectives of forest management with respect to mitigating climate change should be two-fold effort to *protect* and *restore* forests —

- Minimize the release of additional forest carbon into the atmosphere. The best way to *retain* carbon in existing forests is to protect mature and old-growth forests and roadless areas.
- Rebuild depleted carbon stores within forested landscapes. Probably the best way to *rebuild* forest carbon stores in forests is to allow forests that were previously logged or burned to regrow and become mature and old-growth forests.

There are significant complementary benefits of managing forests for carbon storage to ameliorate global climate change. If done carefully, forests managed to provide public services such as clean water, habitat for fish and wildlife, soil conservation, and an enhanced amenity-based economy will also store large amounts of carbon over time.

Forests exhibit a quality known as "ecological inertia" which recognizes that established forests are generally long-lived, resilient to disturbance, and help create conditions suitable for their own survival. This means that our northwest forests may be able to persist through some climate changes and continue to store carbon and provide other benefits, as long as they are not clearcut or severely disturbed. This implies that if we want continued carbon storage in forests that are at the edges of their suitable range we should avoid stand-replacing logging methods (such as clearcutting) and, where ecologically appropriate, we may need to strategically reduce fuels to reduce the risk of stand-replacing fire.

Such fuel reduction must be done carefully however, because excessive removal of vegetation not only compromises carbon storage in both plants and soil, but can also increase fuel loads and fire hazard. Recent fire/fuel models indicate that forest fire hazard can be managed reasonably well by treating about 20-30 per cent of the landscape in strategic locations.

Treating fuel on every acre is neither needed or desired. Logging need not be the primary tool for accomplishing fuel reduction, because non-commercial techniques, such as low-intensity prescribed fire, are available and effective.

Forest Management Recommendations

Private forestlands: Short-rotation clearcutting typically practiced by private industrial forest land-owners is probably the worst possible way to manage forests for carbon storage, because the young forests never develop large carbon stores; significant soil carbon is lost during and after clearcutting, slash disposal, and site preparation; and the resulting wood products produced have limited longevity. Where logging is expected to continue, scientists recommend that carbon release can be mitigated if forest managers:

- Allow trees to grow much longer before harvest (*i.e.*, longer rotations),
- Retain more live trees on every acre during harvest (*i.e.*, thin instead of clearcut),
- Retain more dead wood after harvest (*e.g.* protect snags, practice less intensive slash disposal and site preparation), and
- Take steps to reduce road systems and prevent soil erosion, which would help store more carbon in forest soils.

Public lands: Federal forests can help mitigate climate change if they are restored to their natural-sustainable level of biomass and biodiversity. Large stores of carbon exist within roadless areas and mature and old-growth forests on federal lands. These should be protected from harvest, while previously logged younger forests should be carefully restored to a mature and old-growth condition that has optimal biomass storage. This management approach luckily complements other highly sought-after forest values that are currently under-represented in our forests. Careful management of forests for carbon storage can help resolve ongoing controversies over forestry's impact on water quality, old-growth, roadless areas, fish and wildlife habitat, and scenic values.

Market Solutions: Given humanity's slow response to the growing evidence of human-induced climate change and its consequences, aggressive approaches such as market intervention are now needed. The debate continues on whether a carbon tax or cap-and-trade system is better, but either is better than nothing. A carbon tax system establishes the price of carbon and the market determines how much is sequestered and not emitted. In a cap-and-trade carbon market, government would determine how much total carbon can be emitted from all sources and the market would determine who is allowed to emit the carbon and at what price.

Under current international climate protocols it is possible that forest owners of the Pacific Northwest might seek compensation for storing "extra" carbon. This would reward forest managers for storing carbon that would otherwise be transferred to the atmosphere and help off-set some of the economic costs of managing forests for carbon storage. However, there are unresolved issues about how to account for the full carbon consequences of proposed forest management activities.

For instance, the Kyoto Protocol has some "perverse incentives" that could reward carbon-poor young forests at the expense of carbon-rich old forests, though this is not scientifically supported.

In contrast to the sink management proposed in the Kyoto protocol, which favours young forest stands, we argue that preservation of natural old-growth forests may have a larger effect on the carbon cycle than promotion of regrowth.... [I]ncreasing life-span of the stand, proportionally more carbon can be transferred into a permanent pool of soil carbon (passive soil organic matter or black carbon)... [R]eplacing unmanaged old-growth forest by young Kyoto stands... will lead to massive carbon losses to the atmosphere mainly by replacing a large pool with a minute pool of regrowth and by reducing the flux into a permanent pool of soil organic matter.

Carbon stored in wood products generally do not last as long as they would if left safe inside a mature tree, but we can improve the carbon storage equation by using less wood and by increasing the lifespan of wood products. It's not just American's big cars and SUVs that are a problem; it's also their increasingly large houses. We should consider policies to help reverse the national trend towards larger houses, and we should build houses that last for centuries instead of just decades.

What about Forest Fires?

We cannot avoid the fundamentally dynamic nature of forests. Fire is an unavoidable part of life in western forests and we must stop fighting a losing battle against the inevitable. Most western forests are in some ways *dependent* upon disturbances such as fire, and past fire suppression has exacerbated rather than solved the problem of fire. Our goal should not be to prevent all damage from fires, insects, etc. Fire should be allowed to operate within natural bounds, as long as it doesn't threaten public safety. Communities and property owners in forest settings must take responsibility for becoming fire resilient or fire permeable. We should maintain healthy forest habitat by allowing natural disturbance processes to operate and expect forest carbon stores to ebb and flow, while also allowing forests to grow for long periods (and store lots of carbon) in between these natural disturbances. We must take a long-term and landscape view, so that we *optimize* carbon storage at any given point in space and time in order to *maximize* carbon storage over large landscapes and long time frames.

Fuels could be reduced in forests that are significantly outside the natural range of variability, but this must be done in a strategic and limited way that protects all large fire resilient trees and spatially disconnects large expanses of excessive fuels, while retaining as much biomass as sustainably possible. Current enthusiasm for fuel reduction must be tempered with a realization that removing too much fuel makes forests hotter, dryer, and windier which increases fire hazard and increases decomposition rates, both of which counter carbon storage and other objectives. After fire, the goal should be to retain carbon on site and allow the recovering forest to grow into a mature and old-growth condition. Aggressive replanting as recommended by the timber industry is unsupported because it establishes a dense fuel-laden condition that is susceptible to drought and is soon ripe for another fire. Natural regeneration of forests leads to more diverse and less dense forests, which is preferable from a climate change perspective because the resulting habitat diversity and spatial discontinuity are more resilient to future hazards.

CLIMATE CHANGE COULD AFFECT ENTIRE FOREST ECOSYSTEMS

Off the southern coast of California, the fog rolls in over Santa Cruz Island coating the needles of the Bishop pines with moisture. Water drops form on the needles and roll down to fall to the forest floor, happening over and over on every needle of every tree. The forest ecosystem depends on this fog drip to stay alive. Bishop pines are thought to have proliferated all along the West Coast of the US and Mexico thousands of years ago when the environment was cooler and wetter. Today, the pines are found only on Santa Cruz Island, Santa Rosa Island, an island off Baja California, and in northern California where low-altitude clouds known locally as "June gloom," along with the cooler and wetter climate of the north help keep the trees growing.

A research team from the University of California — Santa Barbara, led by Mariah S. Carbone, studied the influence of clouds on the largest Bishop pine forest of Santa Cruz Island. "When people think about climate change, they're often thinking about temperature and precipitation," said Carbone. "When you think about precipitation, it's rain and snow, depending on where you are. What this study showed is that you can have really important water inputs coming from clouds that influence the carbon cycle."

Clouds are one of the largest uncertainties in global climate, according to Carbone, and the rather large impact they have on forest ecosystems has rarely been studied before, particularly in Southern California. The types of forest ecosystems that grow in coastal California will be altered by changes in cloudiness or cloud height that occur with global warming. While predictions about how California rainfall will change over the next half century are uncertain, scientists believe fog and low-stratus clouds might decline as sea-surface temperatures warm.

Forests such as the relic Bishop pines on Santa Cruz Island and the Sequoias of coastal Northern California remove and store carbon from the atmosphere, making them important in understanding concentrations of greenhouse gases in the future. Additionally, these forests are prized for their aesthetic and recreational qualities.

From May through August, fog is often present in coastal California occurring when the land warms in the spring and summer as moist ocean air is pulled over the cold, upwelling coastal waters. This creates clouds or fog banks as the moist air condenses under a stable atmospheric inversion layer. "The finding that summer fog strongly impacts carbon cycling highlights the need for improved understanding of whether we should expect coastal summer cloud behaviour to change in a warmer world," said A. Park Williams, a former graduate student in UCSB's Geography Department, now at Los Alamos National Laboratory. "A change in summer fogginess could produce temperature, moisture, and carbon feedbacks in coastal ecosystems that easily swamp out the effects expected from increased greenhouse gases alone."

Billions of tons of carbon move between the oceans, lands and atmosphere every year in the carbon cycle, causing increasing amounts of carbon dioxide to be released into the atmosphere. This potent greenhouse gas contributes heavily to climate change. The UCSB team received help collecting information from both NASA and NOAA.

"We used satellite data to map the spatial distribution of cloud cover across the western portion of Santa Cruz Island," said Christopher J. Still, formerly an associate professor at UC Santa Barbara, now at Oregon State University. "We had some evidence of a cloud cover gradient from the coast to inland, based on our own observations and our weather station data, but the satellite data —— from NOAA and NASA satellites —— really allowed us to quantify this gradient across the western half of the island and in particular at our two study sites." This study, linking clouds to soil microbial activity, is unique.

"While most previous research on fog and ecosystems has focused on its role in plant-water relations —— such as the well-known linkages between fog and coast Redwoods —— we show that soil microbial activity and metabolism in these coastal ecosystems may be very dependent on the light but frequent fog drip that occurs during the rainless summer months," said Still. Another unique aspect of the study is the linking of fog and low clouds to the carbon cycle of a coastal ecosystem and how this link impacts tree growth and soil respiration. "What we found was that there are two major effects of the clouds," said Carbone. "One is shading, which keeps the temperatures cooler and the soil moisture higher. The other is fog drip, which is a water input into the soil."

Where the trees were under more cloud cover, the soil moisture content depletes at a slower rate, allowing the trees to grow longer into the dry season. While water pulses from fog drip immediately stimulates microbes in the litter and soil beneath, it does not directly enhance pine tree activity. The microbial activity does, however, release carbon dioxide into the atmosphere.

"This study provides us with a greater understanding of how fog and the low coastal clouds which Californians commonly call 'the marine layer' affect plant growth and the activities of microorganisms in the soil," said John Randall, scientist with The Nature Conservancy. "Their findings of the importance of fog and low clouds in influencing plant growth and survival and microbial metabolism in the soil will help us understand how changes in coastal fogs and low clouds that may accompany climate change will affect the forests and soils on Santa Cruz Island, other islands off the coast of California and Mexico, and all along the coast of the adjacent mainland."

The Nature Conservancy owns and operates Santa Cruz Island in cooperation with the National Park Service and the University of California. From varied landforms supporting more than 600 plant species, to a wide diversity of animal life, to remnants of Native American and European settlements, Santa Cruz Island is an environment rich in biodiversity and cultural resources.

FORESTS OF THE FUTURE BECOME CARBON SOURCES

Just to put the terrestrial biosphere in perspective, there is about ten times more carbon contained in all land plants (plus the soil they grow on) than all the "extra" anthropogenic carbon currently in the atmosphere. Most of the terrestrial carbon is contained in forests which have been significantly depleted by mismanagement.

The question is whether Northwest forests are more likely to store or release carbon under a changing climate. The coupled processes of photosynthesis and respiration/decomposition mirror each other at a global scale to help regulate CO_2 levels and our climate. Photosynthesis captures water and CO_2 and liberates oxygen to create biomass, while respiration consumes biomass and oxygen to liberate CO_2 and water.

Depending on temperature and moisture conditions, among other factors, photosynthesis sometimes dominates leading to net carbon uptake. At other times respiration/decomposition dominate leading to net carbon release. Whether our forests ultimately become net carbon sources or net carbon sinks under the future climate of the Northwest depends on factors that remain uncertain, such as the amount of summer precipitation vs. drought stress, the effects of future climate on fuels and fire hazard, the effects of CO_2 enrichment and climate change on plant physiology, whether forests geographically expand or contract, and whether forests are exploited or protected.

The good news is that slight to moderate warming may increase our forests' ability to store carbon through increased growth and geographic expansion. Pacific Northwest forests might become significant carbon sinks and help mitigate climate change if growing conditions remain favourable and disturbances like fire do not significantly increase. Under warm-wet conditions growing seasons will lengthen, and forest or woodland communities could expand into current rangelands, thus raising the possibility that northwest forests could absorb CO_2 and become a significant net carbon sink.

The bad news is that there is likely a warming *threshold* above which our forests will likely decline due to drought stress and increased disturbances. Drought stress limits the potential photosynthetic benefits of longer growing seasons and CO_2 enrichment. Increasing temperature also increases rates of respiration and decomposition, so under a future climate scenario like this, northwest forests could wither, recede geographically, and become a significant net carbon source. The IPCC tells us that some warming has already occurred and that existing levels of CO_2 already commit us to some additional warming. There is considerable uncertainty about when we may cross the sink to source threshold.

El Niño/Southern Oscillation (ENSO) is a prominent source of multi-year variability in weather and climate around the world. The main signature of ENSO is a periodic (~every 3-8 years) reduction in winds moving westward across the Pacific ocean. This allows warm water to move eastward across the tropical Pacific Ocean. ENSO has strong impacts on ocean nutrient cycling and associated fish populations and birds. ENSO has repercussions far beyond the Pacific ocean, including periodic wide-scale drought in many regions of the world.

Scientists have found a correlation between periodic phenomena like ENSO and years with anomalous global increases in CO_2 which appear to be linked to CO_2 releases from plants, soil, and fire. While there remains debate about this, some have predicted that ENSO may become more frequent and sustained under global warming which could cause a positive feedback favouring respiration over photosynthesis on a global scale.

The source/sink differences could also manifest themselves differently across geography and time periods. "In regions where drought stress is not important because of high levels of precipitation, or if increases in CO_2 concentration increase water use efficiency and thus reduce water stress, longer growing seasons could result in increased growth. Where drought stress is important, a longer growing season may mean only that plant respiration exceeds photosynthesis for a longer time, which would result in reduced growth." So, it is conceivable that moist forests west of the Cascades might remain net carbon sinks, while the dryer forests east of the Cascades might become net sources.

Another study looked at the effects of CO_2 enrichment and climate change on vegetation in the mid- and high-latitudes of the northern hemisphere and found opposing effects in spring and summer. CO_2 uptake was apparently enhanced during warm wet spring season, but looking over the entire growing season, including the dryer summer, CO_2 uptake did not increase. Another paper estimated that western forests might increase in spatial extent while decreasing in their carbon density, *i.e.*, more forested acres, but fewer trees per acre.

FORESTS' EFFECT ON THE CLIMATE

Forests play an important role in regulating the earth's temperature and weather patterns by storing large quantities of carbon and water. This regulatory function has a profound effect on both the local and the global climate. Locally, trees provide shade, which in turn lowers summer temperatures and prevents the soil from drying out, they reduce heat loss from the ground in winter and reduce storm damage by providing shelter from wind. Globally, forests regulate the global carbon cycle, having a profound effect on the climate. As well as this, deforestation is also contributing to climate change. Indeed the CO_2 released each year from forest loss is higher than that released by our yearly transport emissions.

The continued existence of forests is particularly necessary if we are to halt what is known as runaway climate change. Runaway climate change is the point whereby increases in temperature lead to more GHG emissions which in turn leads to increased temperatures. Examples of this include that increased temperatures will melt ice caps which will lead to huge increases in the release of GHGs; and increases in temperature are projected to negatively affect up to two-thirds of existing forests, thereby exacerbating deforestation and increasing the release of carbon.

THE CLIMATE'S EFFECT ON FORESTS

Global warming, which on a geological timescale is occurring in the equivalent of a split second, is significantly disrupting the intricate and poorly understood web of interactions that governs the very structure and composition of forest ecosystems.

This means that around a third of today's forests are likely to change their species composition. A temperature increase of 3°C by 2100 would result in forest ecosystems having to move 500 km towards the poles or 500m in elevation in order to find the same climatic conditions. Such distances are far beyond the average rate of dispersal for individual tree species, let alone entire forest ecosystems.

Early warnings about the consequences of the impacts of climate change on forests have been documented in, among others, The Carbon Bomb: Climate change and the fate of the northern Boreal forests - a 1994 Greenpeace report which states on page 2 that:

- "Studies on the global carbon cycle suggest that boreal forests are

> not absorbing as much carbon as they did before 1976. As a result, the atmosphere already appears to contain 10-15 billion tonnes of carbon more than it would have if forests had continued to absorb carbon at the pre-1976 rate. If boreal forests continue to decline, estimates suggest that burning and rotting of boreal forests could contribute to the release of up to 225 billion tones of extra carbon into the atmosphere, increasing current levels by a third. This would accelerate the rate of climate change."

While it is possible that the boreal forest could expand into the frozen tundra as temperatures increase, such an expansion would likely be delayed by slow tree migration rates. Even in the long-term, the boreal forest is unlikely to move northward fast enough to compensate for the breakdown of boreal forests at the southern part, turning dense forest into open woodlands and grassland, which in turn will result in a lowered biological diversity and a reduced ability of these ecosystems to store carbon and water.

Other forest ecosystems are faced with a similar fate; according to the Intergovernmental Panel on Climate Change(IPCC), a UN panel of climate scientists, it is likely that many tree species will not be able to change their geographic distribution fast enough to keep up with projected shifts in suitable climate and increased rates of extinctions are expected to occur.

What can be Done?

So trees (or the lack of them) are one of the key problems that must be tackled if we are to minimise the effects of climate change. It is not surprising then that many have also suggested that planting trees (carbon sinks) or reducing emissions from deforestation and degradation (REDD) are among the solutions to our climate crisis. SinksWatch believes however that while urgent action is needed to halt forest loss and restored degraded forests, both of these concepts - REDD and carbon sinks - are a dangerous distraction from tackling climate change.

Instead of addressing the root causes of the forest as well as the climate crisis, REDD and carbon sinks are concepts that are linked to a global carbon market where planting trees or reducing deforestation in one place (usually the global South) justifies even more fossil fuel emissions somewhere else (usually in an industrialised country). And many of these carbon sink projects have financed the expansion of large-scale industrial tree plantations.

SinksWatch believes that efforts to tackle forest loss must not justify more fossil fuel emissions because such a trade-off would neither help avert runaway climate change not help save forests in the long run because runaway climate change will also have a major negative impact on the world's forests.

CAPACITIES OF FOREST MANAGERS TO RESPOND TO CLIMATE CHANGE

Many forest managers are aware of climate change and that it may have positive or negative effects on their forests, but only few have a good understanding of the degree of uncertainty of climate change and impact projections.

They are thus susceptible to wrongly interpret those projections unless they receive capacity building on forests and climate change. Most of the existing and proposed actions for mitigation or adaptation to climate change do not require different technical capacities to those already required for SFM. However, it is often not clear to managers that existing best practices may be adequate for the expected or occurring climate induced changes in the natural environment.

Even without considering climate change, many forest managers are not up to date with the best practices for SFM. Preparing for climate change, therefore, needs to include capacity building in SFM practices, including integrated fire management, integrated pest and disease management, harvest planning (including road building), silvicultural practices and the implications actions for nutrient and water cycles.

The practices to be focused upon during capacity building will depend on the country and target group. In the Mediterranean, for example, training for fire management will be more relevant whereas in tropical American pine forest, pests and diseases may be a more relevant topic, although fire management is also important in these areas.

In addition, there is a general need to strengthen the monitoring capacities of forest managers. Awareness raising on adequate monitoring methodologies, tools, and indicators and threshold values that will allow for early detection of changes is also required. This is relevant especially for managers focussing on timber production, forest conservation and protection.

Although the livelihoods of forest dependent people are largely based on NWFPs, there is relatively little awareness of the ecological requirements and management practices of these NWFPs. Capturing that knowledge and sharing it with others will greatly contribute to strengthening adaptive capacity of many communities.

Appropriate Technology

Appropriate technology is needed for three major lines of action: monitoring and research, adaptation and mitigation and will vary according to geographic area, type and objective of forest management activity, scale and intensity of operation and existing local human and financial resources. The major problem with the availability of technology is not its existence, but its location and adaptiveness to local conditions. It may be as simple as having instructions for the use and maintenance of equipment only in a foreign language, not readily understood by the intended users. The cost for acquisition and implementation of this technology is another common problem.

Monitoring and Research

Monitoring and research of vegetation cover usually requires a range of instruments.

Bibliography

Amita Sarkar: *Understanding Invertebrates*, Discovery Publication, Delhi, 2010.

Anil Kumar Thakur and Mithilesh Kumar Sinha: *Climate Change and Environment Management*, Regal Publications, Delhi, 2012.

Ashok Malik: *Dynamics of Forest Ecosystems*, Rajat Publication, Delhi, 2008.

B.K.P. Singh: *Climate Change and Food Safety*, Surendra Publication, Delhi, 2011.

B.S. Bisht, Uma Melkania, Vir Singh, R.K. Srivastava and Nanda Nautiyal: *Climate Change and Agriculture : Emerging Scenarios and Adaptation Strategies*, Biotech Books, Delhi, 2012.

Frank A. Brown: *Invertebrates*, Biotech Books, Delhi, 2002.

G.S. Sandhu and Harshvardhan Bhaskar: *Textbook of Invertebrate Zoology (2 Vols-Set)*, Campus Books, Delhi, 2004.

G.S. Sandhu: *Objective Invertebrate Zoology*, Campus Books International, Delhi, 2005.

G.S.L.H.V. Prasada Rao, G.G.S.N.Rao and V.U.M. Rao: *Climate Change And Agriculture Over India*, PHI Learning, Delhi, 2001.

Golam Kibria, A.K. Yousuf Haroon, Dayanthi Nugegoda and Gavin Rose: *Climate Change and Chemicals : Environmental and Biological Aspects*, New India Publishing Agency, Delhi, 2010.

Gurdarshan Singh: *Manual of Laboratory Specimens : Invertebrates*, Campus Books, Delhi, 2001.

Harsh Vardhan: *Invertebrates, Vols. I and II*, Campus Books, Delhi, 2010.

J. Sundaresan, S. Sreekesh, A.L. Ramanathan, Sonnenschein Leonard and Boojh Ram: *Climate Change and Environment*, Scientific Publication, Delhi, 2013.

K.K. N. Nair, K.V. Bhat, J.K. Sharma and K. Swarupanandan: *Tropical Forest Ecosystem Conservation and Development in South and South-East Asia*, Kerala Forest Research Institute, 1990.

Misten: *Invertebrates*, Ivy Publication, Delhi, 2008.

Nanda Nautiyal, Vir Singh, Neelendra K. Joshi and Akanksha Rastogi: *Climate Change and Biodiversity*, Biotech Books, Delhi, 2012.

Nitish Dogra and Sangeet Srivastava: *Climate Change and Disease Dynamics in India*, Teri Book Store, 2012.

P.L. Gautam, Vir Singh and Uma Melkania: *Ecosystem Diversity and Carbon Sequestration : Climate Change Challenges and A Way Out for Ushering in a Sustainable Future*, Daya Publication, Delhi, 2009.

R.S.K. Barnes, P. Calow, P.J.W. Olive, D.W. Golding and J.I. Spicer: *The Invertebrates : A Synthesis*, Wiley, 2013.

S K Sharma: *Textbook of Invertebrates*, Shree Publication, Delhi, 2003.

S.K. Sharma: *Lower Invertebrate (Non-Chordates)*, JBD Publication, Delhi, 2002.

Tek Bahadur Subba: *Forest Ecosystem in Modern World*, Cyber Tech Publication, Delhi, 2010.

Thomas E Lovejoy and Lee Hannah: *Climate Change and Biodiversity*, TERI, 2006.

V.K. Prabhakar: *Genetic Species and Ecosystem Diversity*, Anmol Publication, Delhi, 2001.

V.P. Singh: *Tropical Forest Ecosystems-Structure and Function*, Scientific Publication, Delhi, 2004.

Index